THE COMBINATION OF
OBSERVATIONS

THE COMBINATION OF OBSERVATIONS

BY

DAVID BRUNT, M.A., B.Sc.

Superintendent of Army Meteorological Services
Meteorological Office, Air Ministry

SECOND EDITION

CAMBRIDGE
AT THE UNIVERSITY PRESS
1931

CAMBRIDGE UNIVERSITY PRESS
Cambridge, New York, Melbourne, Madrid, Cape Town,
Singapore, São Paulo, Delhi, Tokyo, Mexico City

Cambridge University Press
The Edinburgh Building, Cambridge CB2 8RU, UK

Published in the United States of America by Cambridge University Press, New York

www.cambridge.org
Information on this title: www.cambridge.org/9781107697614

First Edition 1917
Reprinted 1923
Second Edition 1931
First paperback edition 2011

A catalogue record for this publication is available from the British Library

ISBN 978-1-107-69761-4 Paperback

PREFACE

THE aim of this book is to give an account of the method of Least Squares, without entering into elaborate descriptions of instruments or of experimental methods. The reader who wishes to apply the method to his own subject can find the experimental details in the textbooks relating to his own special subject, and any attempt to include such details in the present volume would have meant a considerable addition to its size. Hence it has been the author's aim to show how to obtain the best interpretation of the results of experiment, without consideration of the way in which these results are to be obtained.

It cannot be too strongly insisted upon that the methods of Least Squares cannot in any way improve upon the actual observations. The application of these methods to a large number of carelessly conducted experiments cannot in general be expected to yield results as reliable as could be obtained from two or three carefully conducted experiments.

The proof of the Normal Error Law has been based on Hagen's hypotheses regarding errors of observation. In most of the problems of Astronomy, Geodetics, and Physics the errors of observation satisfy the hypotheses, and the application of least square methods is justified. But cases may arise in which particular care is necessary in applying these methods. This is especially true of Biological problems. For organic variability is the resultant of a large number of contributory causes, some of which may have a definite tendency to act always in one direction. The effect of such a bias is to produce an unsymmetrical frequency distribution, and the application of ordinary least square methods is then meaningless. It is thus in no way justifiable to regard Least Squares as a magical instrument applicable to all problems.

Of the arrangement of this volume little need be said. The discussion of methods applicable to problems involving only one unknown quantity will be found in the first four Chapters. Many of the problems of the physicist involve one unknown only, and the first four Chapters contain all the theory that has to be considered in the discussion of such problems. The subject of Chapter VII, the adjustment of conditioned observations, has only been outlined very briefly. The fuller development of the subject, which forms the basis of the adjustment of triangulations, will be found in the works to which reference is made at the end of Chapter VII. The last four Chapters can only be regarded as mere introductions to the subjects discussed, but it was thought that their inclusion in a textbook on Least Squares would be an advantage.

I have to acknowledge my indebtedness to Mr F. J. M. Stratton, of Gonville and Caius College, Cambridge, to whose University lectures I owe most of my knowledge of the subjects discussed in this book, and upon whose notes I have drawn freely. I have also to thank Professor Eddington for many useful suggestions made while the book was in manuscript form, and for permission to extract from his lecture notes a number of interesting examples as well as some portions of the theoretical treatment.

I cannot express adequately my debt to the Cambridge University Press, for the extreme care shown in passing the book through the proof stage. Owing to my being abroad at the time, I was not able to devote as much time as was desirable to the reading of proofs, and but for the unfailing vigilance of the Press, many errors would have been allowed to pass into the text.

DAVID BRUNT.

METEOROLOGICAL SECTION, R.E.
G. H. Q.
November 23, 1916.

TABLE OF CONTENTS

CHAPTER I

ERRORS OF OBSERVATION

CHAPTER II

THE LAW OF ERROR

CHAPTER III

THE CASE OF ONE UNKNOWN

CHAPTER IV

OBSERVATIONS OF DIFFERENT WEIGHT

CHAPTER V

THE GENERAL PROBLEM OF THE ADJUSTMENT OF INDIRECT OBSERVATIONS INVOLVING SEVERAL UNKNOWN QUANTITIES

CHAPTER VI

EVALUATION OF THE MOST PROBABLE VALUES OF THE UNKNOWNS, THEIR WEIGHTS AND PROBABLE ERRORS

CHAPTER VII

THE ADJUSTMENT OF CONDITIONED OBSERVATIONS

CHAPTER VIII

THE REJECTION OF OBSERVATIONS . . . 129

CHAPTER IX

ALTERNATIVES TO THE NORMAL LAW OF ERRORS

CHAPTER X

CORRELATION

CHAPTER XI

HARMONIC ANALYSIS

CHAPTER XII

THE PERIODOGRAM

CHAPTER I

1. THE first serious step in the advance of the sciences which are dependent upon measurements of any kind came with the introduction of instruments. Even the most primitive instruments yielded results far in advance of those obtained by mere estimation with the unaided human senses. With the course of time the development of new instruments has always been in the direction of greater refinement and accuracy. But even the most refined instruments often fail to yield absolutely accurate results; for as a rule it is found that when a series of measurements of the same quantity is taken, the results do not show perfect agreement. The disagreement between individual observations in a series is attributed to errors of observation. Any observation which is of the nature of a measurement is affected by three factors, the instrument used, the external conditions at the time of observation, and the observer. Each of these factors may introduce errors into the observation. We shall consider these three factors in turn.

2. An **instrument** may be defined as a mechanical means of extending the ordinary human faculties, so as to yield measurements of greater refinement than are possible without its aid. The amount of water in a cup, the weight of a stone, the angular distance between two distant points, can be roughly determined by estimation, but measurements of far greater accuracy can be obtained by pouring the water into a graduated vessel, weighing the stone on a balance, and measuring the angular distance between the distant points with a theodolite. The graduated measure, the balance, and the theodolite are examples of instruments.

Just as the graduated vessel is liable to errors of graduation, so the most carefully made instrument is liable to errors of construction, and these errors affect the observations made with the instrument, producing what are known as "instrumental errors." The errors of construction of an instrument may be of the nature of errors of graduation of a scale, periodic errors in a screw, or maladjustment of the separate parts of the instrument. In order to obtain the best possible results, it is necessary to make a careful search for errors in construction, or in actual working, of the instrument, so that their effect may be eliminated. The errors may be of such a nature that it is possible to calculate their effect, and make an empirical correction. But, when possible, it is better to arrange the system of observations in such a way that the instrumental errors eliminate themselves. For example, if measurements have to be made by means of a scale graduated around a circle, it is possible to eliminate most of the errors due to eccentricity of the circle, as well as the greater part of the effect of periodic errors in the graduation of the scale, by placing an even number of verniers, say four or six, at equal distances around the circle*. When it is not possible to eliminate the instrumental errors by the adoption of a suitable method of observation, it may be necessary to carry out a special series of observations for the purpose of calculating empirical corrections for the measurements yielded by the instrument.

The use of refined instruments does not always diminish the difficulties of observation. "The more refined the methods employed, the more vague and elusive does the supposed magnitude become; the judgment flickers and wavers, until at last in a sort of despair *some* result is put down, not in the belief that it is exact, but with the feeling that it is the best we can make of the matter†."

3. The **external conditions** are such parts of the environment of the observer as affect either the observer or his instrument, and are beyond the observer's control; e.g. temperature, wind, or sunlight. If the external conditions be subject to violent change, it may be necessary to suspend work for a time. In some

---

* See Ball, *Spherical Astronomy*, p. 462.

† Lamb, *Presidential Address to Brit. Assoc.* 1904.

cases it is possible to make corrections for the errors introduced by changing external conditions by means of an empirical law determined by an independent set of observations. Such corrections are never perfectly satisfactory, and the observer should avoid the necessity as far as possible, by observing only at times when the external conditions are steady.

4. The **observer** may have personal peculiarities which will affect all his observations. He may always measure an angle as larger (or smaller) than it really is, or he may always tend to note the passage of a star across the wires in the focal plane of an instrument a slight interval before (or after) the true time of transit. A careful and experienced observer appears to commit an error which is generally of the same sign, and approximately of the same magnitude, in a series of similar observations. Such an error is called the "personal equation" of the observer. It can be corrected by comparison with a fixed arbitrary standard. It should be noted, however, that only experienced observers have a well-defined personal equation. An inexperienced observer will commit errors of varying magnitude and sign, and even an experienced observer ceases to have a personal equation when not in his normal state. In order to obtain the best possible results, an observer should not continue to work when he is tired.

From these three sources there will arise errors of a more or less systematic nature, varying according to definite laws with the changing conditions of the observations. When the observed value has been corrected for these sources of error, we might expect the corrected observation to yield the true value of the quantity to be measured. But if a series of observations be made, and corrected in each case for the errors due to the three factors considered above, it will in general be found that the corrected measurements differ among themselves. These individual differences are ascribed to a fourth class of error, known as the accidental error.

5. **Accidental errors** are due to no known cause of systematic or constant error. They are irregular, and more or less unavoidable. The term "accidental" is not used here in its ordinary significance of "chance." Strictly speaking, an observation

of any kind is affected by the state of the whole universe at the time of observation. But as an observer cannot take account of the whole universe and its changes of condition during the time occupied by his observations, he has to limit his attention to those operative causes which he regards as affecting his observations in a measurable degree; i.e. he limits his attention to the "essential conditions. If an observation could be repeated a number of times, and corrected in each case for changes in the essential conditions, the results of all the observations should be identical. But in practice it is found that the individual observations in a series differ among themselves. These differences may be ascribed to the fact that the so-called "essential conditions" do not include all the effective operative causes. There will be other operative causes of error, whose laws of action are unknown, or too complex to be investigated. These causes will introduce errors which will appear to the observer to be accidental.

We can now define accidental errors as errors whose causes and laws of action are unknown. The total accidental error in any individual measurement may be the sum of a number of small accidental errors arising from different causes. Among such errors would be accounted those arising from slight irregular changes in the external conditions, such as the vibration of the image of a distant object on account of air-currents, and the uncertainty of placing a cross-wire upon the image of a scale division; and also irregular changes in the personal equation of the observer. There will also be included in this class the remnants of instrumental errors, but if it should be possible to discover the law of action of any such error, it is thereby removed from the class of accidental errors to the class of systematic errors. Thus, when a distance is measured a large number of times with different parts of a scale, the errors of the scale-division enter into the results in a more or less accidental manner. But if the scale errors be carefully investigated, their effect can be eliminated from the observed values. Carelessness in the handling of an instrument may introduce irregular instrumental errors which fall into the class of accidental errors.

It is thus seen that even when all the systematic errors traceable to the instrument, the external conditions, or the observer, have been corrected, no observation can be regarded as perfect. It will miss perfection on account of the presence of accidental errors. The effect of the accidental errors will differ for different observations in the same series. It is thus impossible to attain certainty in the result of an observation. In practice a series of observations is made, in the hope that the discussion of the series will eliminate the effect of the accidental errors. The problem which we have to solve is that of deciding the best method of conducting this discussion. Our problem may be briefly stated as follows. Given a series of observations, each of which has been made with all possible care, and corrected for all known causes of error, how shall we determine the most probable values of the quantities to be determined? *The material with which the theory deals is supposed to have been cleared of all constant and systematic errors, and to be subject only to accidental errors, whose laws of action are unknown.* The values of the observations, thus corrected, so as to be subject only to accidental errors will in future be referred to as the "observed values."

In what follows, accidental errors will be regarded as obeying the following laws:

(i) A large number of very small accidental errors are present in any observation.

(ii) A positive error and an equal negative error are equally probable.

(iii) The total error cannot exceed a certain reasonably small amount.

(iv) The probability of a small error is greater than the probability of a large error.

As we shall frequently have to refer to constant and systematic errors in the sequel, it will be well to have a clear conception of the meaning of these terms. A **constant error** is one which has the same effect upon all the observations in a series. It has the same magnitude and sign in all the observations. A **systematic error** is one whose sign and magnitude bear a fixed relation to one or more of the conditions of observation. It should be noted that neither of these types of error fulfils the laws of accidental errors given above.

6. Frequency Curves.

Before proceeding further with the theoretical discussion of errors, we shall consider briefly the general nature of the material with which the theory has to deal. The table given below shows the results of a series of observations extracted from the account of the preliminary experiments on photographic transits at the Observatory of Tokio. The first and third columns give the actual observations, while the second and fourth columns give the deviations of the individual observations from the arithmetic mean (4·986) in units of the third decimal place*.

4·974	−12	4·978	−8
82	−4	93	7
78	−8	88	2
89	3	83	−3
93	7	5·001	15
79	−7	5·015	29
84	−2	4·993	7
87	1	91	5
5·001	15	74	−12
4·997	9	71	−15
86	0	67	−19
78	−8	91	5
83	−3	88	2
83	−3	84	−2
90	4	72	−14
91	5	72	−14

The deviations from the arithmetic mean vary from − 19 to + 29.

These deviations (or the actual observations) may be represented graphically as follows (fig. 1). Let the deviations be measured along the horizontal axis, and the number of observations along the vertical axis. Divide the total range of deviation into a number of intervals, say 0 to ± 5, ± 5 to ± 10, etc. For each observation put a dot along the ordinate through the middle point of the interval within which it falls, successive dots on the same ordinate being placed at unit distance apart. The height of the last dot on any ordinate gives the number of observations

---

* These deviations are called the *residuals.*

which fall within the corresponding deviation interval. If the tops of all the ordinates be joined, the resulting broken line represents the frequency of the different measurements, and is called the frequency curve (see figure 1).

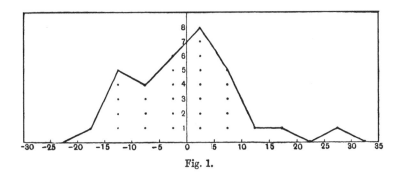

Fig. 1.

Curves obtained by this method, for a moderate number of observations, will generally show one or more peaks, some horizontal portions, and may at some points lie along the horizontal axis. Similar curves will be found in figures 5, 6, 7, and 8. An alternative method of completing the diagram, instead of joining the tops of successive ordinates, is to draw a rectangle whose height is equal to an ordinate, and whose breadth extends over the class-interval, as shown in figure 2. Such a diagram is called a histogram, or a frequency-polygon.

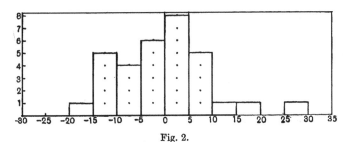

Fig. 2.

Histogram of material shown in figure 1.

If we compare the frequency curves obtained by plotting the repetitions of an observation, using at first a small number of repetitions, and then larger and larger numbers, we find that for a small number of observations the curve shows peaks and valleys and horizontal portions, and changes its form considerably when the number of observations is changed. But as the number of observations becomes larger, the frequency curve tends to approach a fixed form, having in general a single maximum ordinate, while the curve slopes down towards the axis on each side of the maximum. It then becomes convenient to draw a smooth curve through the tops of the ordinates. As the number of observations is still further increased, the successive frequency curves become more and more similar in form, until, when the number of observations is very great, there is no appreciable difference in the form of successive frequency curves. The final form of the curve is that which represents the frequency distribution of an indefinitely large number of observations. This final curve is called the " curve of presumptive errors."

In practice, it is never possible to repeat an observation an indefinitely large number of times, and we have to content ourselves with regarding the frequency curve obtained from a finite number of observations as yielding a reasonably good approximation to the curve of presumptive errors.

The study of a large number of curves of presumptive errors shows a decided similarity in their form, and a strong tendency to approach a typical form distinguished by symmetry about the maximum ordinate. The approach to the typical form is so striking that it is a matter of extreme importance to investigate the possible analytical form of the curve. In the next chapter, starting from certain hypotheses, we shall deduce an analytical formula to represent the typical curve. The purpose of the formula is to express the proportion of the total number of observations whose errors shall lie between any assigned limits, say Δ and $\Delta + d\Delta$; in other words, it will express the probability that the error of a single observation shall lie between Δ and $\Delta + d\Delta$. The use of the words " frequency " and " probability " to denote the same thing is common to writers on this subject. A moment's consideration of the definition of probability will indicate a justification of this custom.

" If, on taking a large number N out of a series of cases in which an event A is in question, A happens on pN occasions, the probability of the event A is said to be p."

The quantity p, so defined, is also the relative frequency of the event A, in a long series of observations.

The law which is commonly held to represent the typical curve of errors is Gauss's Error Law, or the Law of Least Squares, according to which the probability that an observation should have an error between Δ and $\Delta + d\Delta$ is $\dfrac{h}{\sqrt{\pi}} e^{-h^2\Delta^2} d\Delta$, where h is a constant depending on the closeness of the agreement between the observations in the series. This expression will be derived in the next chapter, and its validity will be tested by its application to the adjustment of a number of series of observations. Meanwhile it may be noted that a number of useful results can be derived from the assumption that positive and negative errors are equally probable. Thus the accuracy of the arithmetic mean can be investigated without reference to the actual form of the Law of Errors.

7. The Accuracy of the Arithmetic Mean.

If m_1, m_2, ..., m_n be n determinations of a single unknown quantity, the arithmetic mean is given by

$$a = \frac{1}{n}(m_1 + m_2 + \ldots + m_n),$$

If y_1, y_2, ..., y_n be the errors of the individual determinations, and x the error of the arithmetic mean, then

$$x = \frac{1}{n}(y_1 + y_2 + \ldots + y_n).$$

Squaring each side of this equation, we find

$$x^2 = \frac{1}{n^2}(y_1^2 + y_2^2 + \ldots + y_n^2) + \frac{2}{n^2}(y_1 y_2 + y_1 y_3 + y_2 y_3 + \text{etc.}).$$

The mean value of x^2, say $\bar{x}^2$, is equal to the mean value of the right-hand side of this equation. Since positive and negative errors are equally likely to occur, and all the y's are independent, then in a large number of trials the mean value of $y_r y_s$ will be zero. And since all the n observations considered are supposed to be carried out in precisely the same manner, we may expect the mean values of y_1^2, y_2^2, ..., y_n^2 to be equal. If the mean value of each of the squared terms y_1^2, y_2^2, etc., be μ^2, then

$$\bar{x}^2 = \frac{1}{n^2}(n\mu^2) = \frac{\mu^2}{n}$$

and

$$\bar{x} = \frac{\mu}{\sqrt{n}}.$$

Hence it follows that the accuracy of the arithmetic mean is $\sqrt{n}$ times the accuracy of a single observation; a result of fundamental importance in the theory of errors.

CHAPTER II

8. Hagen's Proof of Gauss's Error Law*.

Hagen based his proof on the assumption that an accidental error consists of the algebraic sum of a very large number of infinitesimal errors, all of equal magnitude, and as likely to be positive as negative. It has already been suggested that the accidental error occurring in any one observation may be composed of a number of errors due to slight changes in the external conditions, remnants of instrumental errors, or irregularities in the individual peculiarities of the observer. Each of these components may in turn be regarded as due to a large number of elementary causes. And so Hagen's hypothesis is in no way a violation of our knowledge of the nature of accidental errors.

Let the total number of elementary errors be n, where n is a number to which we can assign no limit. If the magnitude of each of the elemental errors be ϵ, then it will be possible for an error as great as $n\epsilon$ to occur, in the extreme case where all the small errors occur with the same sign. An error $(n - 2m)\,\epsilon$ will occur when $n - m$ of the elemental errors occur with a positive sign, and the remaining m with a negative sign. The number of ways in which this can happen is simply the number of ways in which we can select m out of the n errors. The selected m errors will have one sign, and the remaining $n - m$ will have the opposite sign. The selection can be made in

$$\frac{n\,!}{n - m\,!\; m\,!} \text{ ways.}$$

* Hagen, *Gründzuge der Wahrscheinlichkeitsrechnung* (Berlin, 1837).

Let $x - \epsilon$ be the error due to $n - m$ errors $+ \epsilon$, and m errors $- \epsilon$.

Then
$$x - \epsilon = (n - 2m)\,\epsilon$$

or
$$x = (n - 2m + 1)\,\epsilon.$$

If $f(x - \epsilon)$ be the frequency of the error $x - \epsilon$,
$$f(x - \epsilon) = \frac{n\,!}{n - m\,!\; m\,!}$$

An error $x + \epsilon$, or $(n - 2m + 2)\,\epsilon$, will be formed by $n - m + 1$ elemental errors $+ \epsilon$, and $m - 1$ errors $- \epsilon$. Hence
$$f(x + \epsilon) = \frac{n!}{n - m + 1!\; m - 1!}.$$

It follows that
$$\frac{f(x + \epsilon)}{f(x - \epsilon)} = \frac{m}{n - m + 1}.$$

Since the elemental errors ϵ are supposed to be infinitesimally small in comparison with the finite composite errors of actual observations, the latter may be assumed to vary continuously from $-n\epsilon$ to $+n\epsilon$, the extreme error $\pm n\epsilon$, whose relative frequency is very small, being assumed to be infinite. If we define $\phi(x)\,dx$ as the proportion of errors between $x - \tfrac{1}{2}dx$ and $x + \tfrac{1}{2}dx$, $\phi(x)$ may be regarded as a continuous function. But the function $f(x)$ gives the frequency of the errors x, where all the possible values of x are separated by intervals of 2ϵ. We may bring the function $\phi(x)$ and $f(x)$ into line by regarding $f(x)$ as the frequency of all errors between $x - \epsilon$ and $x + \epsilon$. We then have the equation
$$f(x) = C\phi(x)\,.\,2\epsilon.$$

It follows that
$$\frac{\phi(x + \epsilon)}{\phi(x - \epsilon)} = \frac{f(x + \epsilon)}{f(x - \epsilon)} = \frac{m}{n - m + 1},$$
$$\frac{\phi(x + \epsilon) - \phi(x - \epsilon)}{\phi(x + \epsilon) + \phi(x - \epsilon)} = -\frac{(n - 2m + 1)}{n + 1},$$

Neglecting squares of ϵ, we may write
$$\phi(x + \epsilon) = \phi(x) + \epsilon\,\frac{d\phi}{dx},$$
$$\phi(x - \epsilon) = \phi(x) - \epsilon\,\frac{d\phi}{dx}.$$

Substituting these values in the above equation, we find

$$\frac{2\epsilon \dfrac{d\phi}{dx}}{2\phi(x)} = -\frac{n-2m+1}{n+1} = -\frac{x}{(n+1)\epsilon},$$

or

$$\frac{1}{\phi} \cdot \frac{d\phi}{dx} = -\frac{x}{(n+1)\epsilon^2}.$$

Integrating this equation, we find

$$\log \phi(x) = -\frac{x^2}{2(n+1)\epsilon^2} + \text{const.}$$

$$= -h^2 x^2 + \text{const.}$$

where

$$h^2 = \frac{1}{2(n+1)\epsilon^2},$$

$$\phi(x) = A e^{-h^2 x^2}.$$

The constant h^2 depends upon the nature of the errors entering into the observations. The value of the constant A is easily derived. $\phi(x) \cdot dx$ gives the proportion of errors between the limits $x - \frac{1}{2}dx$ and $x + \frac{1}{2}dx$, or, practically, between the limits x and $x + dx$. Since all the possible errors must lie between $-\infty$ and $+\infty$, it follows that

$$\int_{-\infty}^{+\infty} \phi(x)\, dx = 1 = A \int_{-\infty}^{+\infty} e^{-h^2 x^2}\, dx$$

$$= 2A \int_{0}^{\infty} e^{-h^2 x^2}\, dx$$

$$= 2A \times \frac{\sqrt{\pi}}{2h}.$$

$$\therefore \quad A = \frac{h}{\sqrt{\pi}}.$$

Finally, we may write

$$\phi(x) = \frac{h}{\sqrt{\pi}} e^{-h^2 x^2}.$$

This is the functional form of Gauss's Error Law, or, as it is sometimes called, the Normal Error Law. Its interpretation is that the relative number of observations in a series, whose errors lie within the limits x and $x + dx$, is

$$\frac{h}{\sqrt{\pi}} e^{-h^2 x^2}\, dx.$$

In a long series of observations, n in number, the number of observations whose errors lie between x and $x + dx$ is

$$\frac{nh}{\sqrt{\pi}}\, e^{-h^2 x^2}\, dx.$$

9. Herschel's Proof of the Error Law.

Consider the distribution of shots fired at a target. Let axes of coordinates be drawn through the centre of the target, the x-axis horizontal, the y-axis vertical. Let P, (x, y), be the mark of one shot on the target. Then x and y are components of the error of placing the point P. Since the shot is as likely to go to the right as to the left of the centre of the target, the probability of an error between x and $x + dx$ in the x-coordinate is of the form $\phi(x^2)\,.\,dx$. Similarly the probability of an error between y and $y + dy$ in the other coordinate is of the form $\phi'(y^2)\,.\,dy$. We shall assume that the functions ϕ and ϕ' are of the same form. This is equivalent to assuming that a large number of shots fired at the target would be distributed indiscriminately about the centre, showing no special symmetry of distribution about the horizontal and vertical axes, as opposed to any other axis through the centre of the target.

Then the probability that the point P should have coordinates lying between x and $x + dx$, and between y and $y + dy$, respectively, will be

$$\phi(x^2)\,.\,\phi(y^2)\,.\,dx\,.\,dy.$$

In other words, the probability that the point P should lie within a given small region of area dA is

$$\phi(x^2)\,.\,\phi(y^2)\ \ dA,$$

x and y being the coordinates of any point within the area dA.

But if another pair of axes Ox', Oy' be drawn through the centre of the target, so that Ox' passes through the area dA, the probability that a single shot should be placed within the area dA can also be expressed by

$$\phi(x'^2)\,.\,\phi(y'^2)\,dA,$$

or

$$\phi(x^2 + y^2)\,.\,\phi(0)\,dA.$$

$$\therefore\ \ \phi(x^2)\,.\,\phi(y^2) = \phi(x^2 + y^2)\,.\,\phi(0).$$

This is a functional equation whose solution is

$$\phi(x^2) = A e^{kx^2}.$$

Since small errors are more probable than large errors, $\phi(x^2)$ must decrease as x increases, and k must therefore be negative. Putting $k = -h^2$, we can show as before that

$$A = \frac{h}{\sqrt{\pi}}.$$

Finally we may write

$$\phi(x^2) = \frac{h}{\sqrt{\pi}} e^{-h^2 x^2},$$

which agrees with the form of the error law derived above (§ 8).

10. A Generalised Form of Hagen's Proof *.

Suppose that the error of observation is made up of a large number of independent elementary errors, and that the probability that any one of these elementary errors shall lie between ϵ and $\epsilon + d\epsilon$ is $g(\epsilon) d\epsilon$. We shall assume that positive and negative errors are equally likely, so that $g(\epsilon)$ is an even function, but otherwise $g(\epsilon)$ may be regarded as being quite arbitrary. Then

$$\left.\begin{array}{r}\displaystyle\int_{-\infty}^{+\infty} g(\epsilon)\, d\epsilon = 1, \\[2mm] \displaystyle\int_{-\infty}^{+\infty} \epsilon\, g(\epsilon)\, d\epsilon = 0. \\[2mm] \displaystyle\int_{-\infty}^{+\infty} \epsilon^2\, g(\epsilon)\, d\epsilon = k^2. \end{array}\right\} \quad \dots\dots\dots\dots\dots(1).$$

Further let

The quantity k so defined may be called the mean square elementary error.

Let

$$f(n, x)\,.\,dx$$

be the probability that the resultant error due to n of the elementary errors lies between x and $x + dx$. Then we have

$$f(n+1, x) = \int_{-\infty}^{+\infty} f(n, x - \epsilon)\,.\,g(\epsilon)\, d\epsilon.$$

This equation expresses the fact that the probability that $n + 1$ errors add up to x, is made up of the probability that the first

---

* The above proof is due to Professor Eddington.

n errors add up to $x - \epsilon$, multiplied by the probability that the remaining error is ϵ, summed for all possible values of ϵ.

Expand both sides of the equation by Taylor's Theorem *.

$$f(n,\, x) + \frac{\partial}{\partial n} \cdot f(n,\, x) + \,..$$

$$= \int_{-\infty}^{+\infty} \left\{ f(n,\, x) - \epsilon \frac{\partial}{\partial x} f(n,\, x) + \tfrac{1}{2} \epsilon^2 \frac{\partial^2}{\partial x^2} f(n,\, x) - \text{etc.} \right\} g\,(\epsilon)\, d\epsilon$$

$$= f(n,\, x) + \tfrac{1}{2} k^2 \frac{\partial^2}{\partial x^2} f(n,\, x) + \ldots \text{ by (1).}$$

Write $nk^2 = t$. Then $\sqrt{t}$ is the mean square error of the observations, and remains finite when n is taken infinitely great, and k^2 accordingly infinitely small.

We then have

$$f + k^2 \frac{\partial f}{\partial t} + \ldots = f + \tfrac{1}{2} k^2 \frac{\partial^2 f}{\partial x^2} + \ldots,$$

the omitted terms being of the order of k^4, when k is small. Taking now a very large number of very small elementary errors, so that k^4 is negligible in comparison with k^2, this gives

$$\frac{\partial f}{\partial t} = \tfrac{1}{2} \frac{\partial^2 f}{\partial x^2} \quad \ldots\ldots\ldots\ldots\ldots\ldots\ldots(2).$$

It is easily verified that a solution of this equation is

$$f = C t^{-\frac{1}{2}} e^{-\frac{x^2}{2t}} \quad \ldots\ldots\ldots\ldots\ldots\ldots(3).$$

It remains to show that this is the solution applicable to our problem. The differential equation (2) is the same as that which determines the conduction of heat along a bar. Now when the mean square error of an observation is zero, the probability of an error x vanishes except for $x = 0$; but unit probability is then concentrated into an infinitesimal range at $x = 0$. The corresponding condition in the heat problem is that initially ($t = 0$) the bar is everywhere at zero temperature except that a unit quantity of heat is concentrated in it at the point $x = 0$. Also, whatever be the mean square error, f must vanish at $x = +\infty$ and $x = -\infty$; in the heat problem this means that the temperature at the two infinitely distant ends is zero. Now it is known from the theory

---

* Difficulties as to the possible divergence of the Taylor expansion for large values of ϵ may be avoided by introducing the additional assumption that $g\,(\epsilon) = 0$ for values of ϵ beyond certain limits.

of the conduction of heat that these two conditions—the initial condition and the end condition—are sufficient to determine *uniquely* a solution of the equation. Hence if we can obtain a solution for f satisfying (1) $f = 0$ when $t = 0$, for all values of x except $x = 0$, and (2) $f = 0$ when $x = +\infty$, and $x = -\infty$, for all values of t, this will be the only possible solution for the error law. It is easily seen that the solution (3) does satisfy these conditions.

The constant C must be chosen so that

$$\int_{-\infty}^{+\infty} f \,.\, dx = 1,$$

and writing

$$h^2 = \frac{1}{2t},$$

we obtain the expression in the usual form,

$$f = \frac{h}{\sqrt{\pi}}\, e^{-h^2 x^2}.$$

11. The proofs of the Normal Error Law given above are based on certain definite hypotheses concerning the nature of accidental errors. It has been shown that, if the accidental errors to which a series of observations is liable satisfy these hypotheses, the errors of observation will be distributed according to the normal law. The final justification of the use of Gauss's Error Curve rests upon the fact that it works well in practice, and yields curves which in very many cases agree very closely with the observed frequency curves. The normal law is to be regarded as *proved by experiment,* and *explained* by Hagen's hypothesis. When the curve of frequency of the actual errors is not of the form of the normal curve, we may safely conclude that the nature of the accidental errors concerned is not in accordance with Hagen's hypothesis.

The normal curve has applications in a region where deviations from a mean value are considered, though these deviations are not, properly speaking, of the same nature as accidental errors of observation; e.g., it is frequently applied in biological questions (*vide* Ex. 3, p. 42). The use of the normal curve in such cases is justified only when the differences between individual cases are produced by causes whose mode of action is in accordance with Hagen's hypothesis.

12. The Form of the Error Curve.

The equation of the Error Curve is

$$y = \frac{h}{\sqrt{\pi}} \, e^{-h^2 x^2}.$$

The curve is symmetrical about the axis of y. The maximum ordinate occurs at $x = 0$, and has the value $\frac{h}{\sqrt{\pi}}$. The curve can easily be constructed by the use of tables of logarithms.

Differentiating this equation twice we obtain

$$\frac{d^2 y}{dx^2} = \frac{2h^3}{\sqrt{\pi}} \left(2h^2 x^2 - 1\right) e^{-h^2 x^2},$$

showing that the curve has points of inflexion at

$$x = \pm \frac{1}{\sqrt{2}h}.$$

The form of the curve is shown in figure 3.

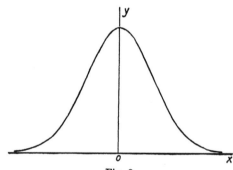

Fig. 3.

Gauss's Error Curve.

The probability that the error of an observation shall lie between a and b is

$$\frac{h}{\sqrt{\pi}} \int_a^b e^{-h^2 x^2} \, dx, \quad \text{or} \quad \frac{1}{\sqrt{\pi}} \int_{ha}^{hb} e^{-t^2} \, dt.$$

The value of this integral can be derived from the table below. The table gives the value of $\Theta(t)$, which is equal to

$$\frac{2}{\sqrt{\pi}} \int_0^{\rho x} e^{-t^2} \, dt,$$

for a series of values of x from 0 to 5·0, ρ being a constant whose value is 0·47694, the reason for the use of which will appear later.

$$\text{TABLE OF } \Theta(t) = \frac{2}{\sqrt{\pi}} \int_0^{\rho x} e^{-t^2} dt.$$

x	$\Theta(t)$	x	$\Theta(t)$
0·0	0·000	2·6	0·921
0·1	0·054	2·7	·931
0·2	·107	2·8	·941
0·3	·160	2·9	·950
0·4	·213	3·0	·957
0·5	·264	3·1	·963
0·6	·314	3·2	·969
0·7	·363	3·3	·974
0·8	·411	3·4	·978
0·9	·456	3·5	·982
1·0	·500	3·6	·985
1·1	·542	3·7	·987
1·2	·582	3·8	·990
1·3	·619	3·9	·992
1·4	·655	4·0	·993
1·5	·688	4·1	·994
1·6	·719	4·2	·995
1·7	·748	4·3	·996
1·8	·775	4·4	·997
1·9	·800	4·5	·998
2·0	·823	4·6	·998
2·1	·843	4·7	·998
2·2	·862	4·8	·999
2·3	·879	4·9	·999
2·4	·895	5·0	·999
2·5	·908	∞	1·000

A table of $\dfrac{2}{\sqrt{\pi}} \displaystyle\int_0^t e^{-t^2} dt$, or Erf (t), is given in Appendix II.

For small values of x, the integral $\displaystyle\int_0^x e^{-x^2} dx$ may be obtained by expanding e^{-x^2}.

$$e^{-x^2} = 1 - x^2 + \frac{1}{2} x^4 - \frac{1}{1.2.3} x^6 + \text{etc.}$$

This series is uniformly convergent, and may therefore be integrated term by term.

$$\int_0^x e^{-x^2} dx = x - \frac{x^3}{1!\,3} + \frac{x^5}{2!\,5} - \frac{x^7}{3!\,7} + \text{etc.}$$

When x is not small the above series converges very slowly, and it is then better to use another formula obtained by integrating by parts.

$$\int e^{-x^2}\, dx = -\frac{1}{2x}\, e^{-x^2} - \frac{1}{2}\int \frac{1}{x^2}\, e^{-x^2}\, dx$$

$$= -\frac{1}{2x}\, e^{-x^2} + \frac{1}{2x^3}\, e^{-x^2} + \frac{1.3}{2^2}\int \frac{1}{x^4}\, e^{-x^2}\, dx.$$

Continuing the process we find

$$\int_x^\infty e^{-x^2}\, dx = \frac{e^{-x^2}}{2x}\left\{1 - \frac{1}{2x^2} + \frac{1.3}{(2x^2)^2} - \frac{1.3.5}{(2x^2)^3} + \text{etc.}\right\},$$

and since

$$\int_0^x e^{-x^2}\, dx = \int_0^\infty e^{-x^2}\, dx - \int_x^\infty e^{-x^2}\, dx = \frac{\sqrt{\pi}}{2} - \int_x^\infty e^{-x^2}\, dx,$$

we can apply the second series to evaluate $\int_0^x e^{-x^2}\, dx$.

When x is large a still better method is to use a series due to Schlomilch*.

$$\int_x^\infty e^{-x^2}\, dx = \frac{e^{-x^2}}{2x}\left\{1 - \frac{1}{2\,(x^2+1)} + \frac{1}{2^2\,(x^2+1)\,(x^2+2)}\right.$$

$$- \frac{5}{2^3\,(x^2+1)\,(x^2+2)\,(x^2+3)} + \frac{9}{2^4\,(x^2+1)\,(x^2+2)\,(x^2+3)\,(x^2+4)}$$

$$\left.- \frac{129}{2^5\,(x^2+1)\,(x^2+2)\,(x^2+3)\,(x^2+4)\,(x^2+5)} + \ldots \text{etc.}\right\}.$$

This series converges very rapidly for large values of x.

The error curve drawn above, represented by the function

$$y = \frac{h}{\sqrt{\pi}}\, e^{-h^2 x^2},$$

extends to infinity along the axis of x, in both directions, having the axis of x as an asymptote. In a system of errors accurately represented by this curve, the errors are continuous, and extend from $-\infty$ to $+\infty$. In actual observations the errors are never greater than a reasonably small finite limit, and the curve which would accurately represent the system of errors should meet the x-axis at a finite distance from the origin. But as the normal

---

* *Kompendium der höhern Analysis*, Bd. II, p. 266.

curve of errors shown in figure 3 rapidly approaches the axis of x, so that its ordinates become extremely small at a short distance from the origin, no considerable error is introduced by regarding the curve of actual errors as approaching the axis of x asymptotically, and so regarding the limits of possible errors of measurement as being extended to $\pm \infty$.

13. The Arithmetic Mean.

If $x_1, x_2, ..., x_n$ be n observed values of a quantity x, the errors of the separate observations are $x_1 - x$, $x_2 - x$, etc. The probability of making this system of errors is proportional to

$$e^{-h^2 \Sigma (x_r - x)^2}.$$

The value of x must be so chosen that this probability shall be a maximum, or so that $\Sigma (x_r - x)^2$ shall be a minimum. Differentiating the last expression we obtain an equation for x,

$$\Sigma (x_r - x) = 0, \quad \text{or} \quad \Sigma x_r - nx = 0.$$

Hence

$$x = \frac{1}{n} \Sigma x_r = \frac{1}{n} (x_1 + x_2 + ... + x_n) \quad..............(1).$$

The most probable value of the unknown is therefore the arithmetic mean of the observed values.

This result may be obtained without the aid of the differential calculus. If $\bar{x}$ be the arithmetic mean of $x_1, x_2, ..., x_n$, and x any other value of the unknown,

$$\Sigma (x_r - x)^2 - \Sigma (x_r - \bar{x})^2 = n (x^2 - \bar{x}^2) - 2 (x - \bar{x}) \Sigma x_r$$
$$= n (x^2 - \bar{x}^2) - 2n\bar{x} (x - \bar{x})$$
$$= n (x - \bar{x})^2.$$
$$\therefore \ \Sigma (x_r - x)^2 = \Sigma (x_r - \bar{x})^2 + n (x - \bar{x})^2 \quad.........(2).$$

It follows that $\Sigma (x_r - x)^2$ is least when $x = \bar{x}$, or when x coincides with the arithmetic mean of the observed values.

It will be convenient to use the abbreviation A.M. to denote the arithmetic mean.

The quantities obtained by subtracting the A.M. $\bar{x}$ from each of the observed values are called the residuals, and are usually denoted by v_1 v_2, etc. We thus have a set of n equations,

$$x_1 - \bar{x} = v_1,$$
$$x_2 - \bar{x} = v_2,$$
$$............$$
$$x_n - \bar{x} = v_n.$$

Also $\quad v_1 + v_2 + \ldots + v_n = (x_1 + x_2 + \ldots + x_n) - n\bar{x} = 0 \quad \ldots \ldots (3)$.

The name of the method of "Least Squares" is due to the fact brought out by equation (2), that the most probable value of the unknown is that value for which the sum of the squares of the residuals is least.

In actual computation it is necessary to have numerical checks upon the calculated values of the A.M. and the sum of the squares of the residuals. If $\bar{x}$ be the A.M. and x' any other quantity, then

$$\bar{x} = x' + \frac{\Sigma (x_r - x')}{n},$$

$$\Sigma (x_r - \bar{x})^2 = - n (\bar{x} - x')^2 + \Sigma (x_r - x')^2.$$

A calculation from some second base x' thus affords a useful check upon the value of the mean and upon the sum of the squares of the residuals. Also, in some cases, it may be more convenient to evaluate $\Sigma (x_r - x')^2$ than $\Sigma (x_r - \bar{x})^2$.

The accuracy of the A.M. can also be checked by means of the sum of the residuals, since $\Sigma v = 0$.

14. Proof of the Normal Law from the Principle of the Arithmetic Mean.

In § 13 above we derived the principle of the arithmetic mean from the Normal Law. It is interesting to note that we can invert the process, and derive the Normal Law from the principle of the arithmetic mean.

Let the probability that the error of an observation should lie between Δ and $\Delta + d\Delta$ be represented by $\phi (\Delta) d\Delta$. Our problem is to find the form of the function $\phi(\Delta)$. As $d\Delta$ tends to zero $\phi (\Delta) d\Delta$ also tends to zero; or in other words, the probability of making an error of the exact value Δ is zero. When we speak of the "probability of an error Δ," we shall interpret this expression as meaning the probability of an error between $\Delta - \alpha$ and $\Delta + \alpha$, where α is a small quantity which is just inappreciable in the observation in question. With this convention we may say that the probability of an error Δ is $C\phi (\Delta)$, where C is a constant.

Let $x_1, x_2, \ldots, x_n$ be n observed values of an unknown quantity x. If x be the assumed value of the unknown, the errors of the separate observations will be $\Delta_1, \Delta_2, ..., \Delta_n$, defined by

$$\Delta_1 = x_1 - x, \quad \Delta_2 = x_2 - x, \text{ etc.}$$

The probability of the occurrence of an error Δ_r is $C\phi(\Delta_r)$, and the probability of the occurrence of the system of errors $\Delta_1, \Delta_2, ..., \Delta_n$ is

$$C^n \phi(\Delta_1) \phi(\Delta_2) ... \phi(\Delta_n).$$

The assumed value of the unknown, x, must be the most probable value of the unknown; in other words, it will be such as to make the system of errors Δ_1, Δ_2, etc., the most probable. It follows that the probability

$$C^n \phi(\Delta_1) \phi(\Delta_2) ... \phi(\Delta_n)$$

must be a maximum for the assumed value of x. Differentiating with respect to x, we may write this condition in the form

$$\frac{\phi'(\Delta_1)}{\phi(\Delta_1)} \frac{d\Delta_1}{dx} + \frac{\phi'(\Delta_2)}{\phi(\Delta_2)} \frac{d\Delta_2}{dx} + ... + \frac{\phi'(\Delta_n)}{\phi(\Delta_n)} \frac{d\Delta_n}{dx} = 0,$$

where
$$\phi'(\Delta_r) = \frac{d}{dr} \phi(\Delta_r).$$

But since
$$\Delta_1 = x_1 - x, \quad \Delta_2 = x_2 - x, \text{ etc.},$$

$$\frac{d\Delta_1}{dx} = \frac{d\Delta_2}{dx} = \text{etc.} = -1,$$

and the condition for a maximum may be written

$$\Sigma \frac{\phi'(\Delta_r)}{\phi(\Delta_r)} = 0,$$

or
$$\Sigma \frac{\phi'(\Delta_r)}{\Delta_r \phi(\Delta_r)} \Delta_r = 0 ....(1).$$

Equation (1) must be satisfied for the most probable value of x.

We now assume that this value of x is the A.M. With this assumption, we may write down the additional equation

$$\Sigma \Delta_r = 0 \qquad ....(2),$$

which is simply equation (3) of § 13.

Equations (1) and (2) must be simultaneously satisfied by the same value of the unknown, and so they must be identical.

$$\therefore \quad \frac{\phi'(\Delta_1)}{\Delta_1 \phi(\Delta_1)} = \frac{\phi'(\Delta_2)}{\Delta_2 \phi(\Delta_2)} = ... = \frac{\phi'(\Delta_n)}{\Delta_n \phi(\Delta_n)}$$

$$= \text{constant} = k \text{ say.}$$

Then
$$\frac{\phi'(\Delta)}{\Delta\phi(\Delta)} = k \quad \text{or} \quad \frac{\phi'(\Delta)}{\phi(\Delta)} = k\Delta$$

yields a differential equation whose solution determines the form of $\phi(\Delta)$. Integrating the equation, we find

$$\log \phi(\Delta) = \tfrac{1}{2}k\Delta^2 + \text{const.}$$

or
$$\phi(\Delta) = Ae^{\frac{1}{2}k\Delta^2}.$$

The product $\phi(\Delta_1)\,\phi(\Delta_2)\ldots\phi(\Delta_n)$ must be a maximum for the assumed value of x. Hence the series $\Sigma \log \phi(\Delta_r)$ must be a maximum. The condition to be satisfied is that

$$\frac{d^2}{dx^2}\,\Sigma \log \phi(\Delta) \text{ shall be negative,}$$

or
$$\Sigma \frac{d^2}{dx^2}\,(\tfrac{1}{2}k\Delta^2) \text{ shall be negative.}$$

Since
$$\Delta_r = x_r - x,$$

$$\frac{d^2}{dx^2}\,\Delta_r{}^2 = 2,$$

and
$$\Sigma \frac{d^2}{dx^2}\,(\tfrac{1}{2}k\Delta_r{}^2) = n \cdot k.$$

It follows that k must be negative.

Putting $\tfrac{1}{2}k = -h^2$, we may write

$$\phi(\Delta) = Ae^{-h^2\Delta^2}.$$

The value of A may be derived as in § 8, yielding

$$\phi(\Delta) = \frac{h}{\sqrt{\pi}}\,e^{-h^2\Delta^2}.$$

15.   The Law of Error of a linear function of two independent quantities whose laws of error are known.

Let m_1, m_2 be two *independent* observed quantities, obeying the error laws

$$\frac{h_1}{\sqrt{\pi}}\,e^{-h_1{}^2x^2} \quad \text{and} \quad \frac{h_2}{\sqrt{\pi}}\,e^{-h_2{}^2x^2}, \text{ respectively}.$$

The probability of the occurrence of an error between x_1 and $x_1 + dx_1$ in m_1 is

$$\frac{h_1}{\sqrt{\pi}}\,e^{-h_1{}^2x_1{}^2}\,dx_1,$$

and the probability of the occurrence of an error between x_2 and $x_2 + dx_2$ in m_2 is

$$\frac{h_2}{\sqrt{\pi}} e^{-h_2^2 x_2^2} dx_2.$$

Since m_1 and m_2 are independent, the probability of the simultaneous occurrence of these errors in m_1 and m_2 is the product of the two separate probabilities. Calling this probability p, we have the equation

$$p = \frac{h_1 h_2}{\pi} e^{-h_1^2 x_1^2 - h_2^2 x_2^2} dx_1 dx_2.$$

If the linear function in question be

$$a_1 m_1 + a_2 m_2 = F,$$

the corresponding error will lie between

$$a_1 x_1 + a_2 x_2 = X \text{ (say)}$$

and $\qquad a_1 (x_1 + dx_1) + a_2 (x_2 + dx_2) = X + dX.$

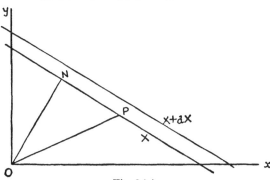

Fig. 3 (a).

Let $\qquad\qquad h_1 x_1 = x, \quad h_2 x_2 = y.$

Then $\qquad\qquad X = \dfrac{a_1}{h_1} x + \dfrac{a_2}{h_2} y.$

The probability of the simultaneous occurrence of errors between x and $x + dx$ and between y and $y + dy$ is

$$p = \frac{h_1 h_2}{\pi} e^{-(x^2 + y^2)} \frac{dx \, dy}{h_1 h_2} = \frac{1}{\pi} e^{-\lambda(x^2 + y^2)} dx \, dy.$$

The probability that the error in F lies between X and $X + dX$ is $\int p$ over the area between the lines X and $X + dX$. In figure 3 (a)

N is the foot of the perpendicular drawn from the origin on the line

$$\frac{a_1}{h_1}x + \frac{a_2}{h_2}y = X,$$

$$x^2 + y^2 = ON^2 + PN^2$$

$$= \frac{X^2}{\left(\frac{a_1}{h_1}\right)^2 + \left(\frac{a_2}{h_2}\right)^2} + r^2,$$

where $r = PN$.

Let

$$\frac{1}{h^2} = \frac{a_1^2}{h_1^2} + \frac{a_2^2}{h_2^2}.$$

Then $ON = hX$, and the distance between the lines X and $X + dX$ is $h\,dX$, the difference between the perpendiculars from the origin on these two lines.

The probability that the error in F lies between X and $X + dX$

$$= \frac{1}{\pi}e^{-h^2X^2}\int_0^\infty e^{-r^2}\,dr\,h\,dX$$

$$= \frac{h}{\sqrt{\pi}}e^{-h^2X^2}\,dX.$$

Thus the law of error of F is of the same form as the laws of error of m_1 and m_2, the parameter h being determined by the equation given above. The result may be generalised for any number of variables, adding one at a time to F, so that if

$$F = a_1 m_1 + a_2 m_2 + a_3 m_3 + \ldots + a_n m_n,$$

and $h, h_1, h_2, \ldots, h_n$ are the parameters of $F, m_1, m_2, \ldots, m_n$, then

$$\frac{1}{h^2} = \frac{a_1^2}{h_1^2} + \frac{a_2^2}{h_2^2} + \ldots + \frac{a_n^2}{h_n^2} = \Sigma\frac{a^2}{h^2}.$$

For three variables, these results are readily derived by the geometrical method used above for two variables. This is left to the reader as an easy exercise in the application of the methods. The most general case for n variables can also be derived in the same way, using the geometry of n-dimensional space, in which the volume of a sphere of radius R is $\dfrac{\pi^{\frac{1}{2}n}R^n}{\Gamma(\frac{1}{2}n+1)}$, the denominator being the ordinary Gamma Function.

16. The Median.

The median, which is the value of the unknown which has as many observed values on one side of it as on the other, is the natural competitor of the arithmetic mean, and it is interesting to consider the law of error which follows from the assumption that the median is the most probable value of the unknown.

Let x' be the median, and let $f(x-x')\,dx$ be the probability of making an observation between x and $x+dx$. If $x_1, x_2, ..., x_n$ be the observed values, we have to make the product

$$f(x_1-x')f(x_2-x')\,...f(x_n-x') \text{ a maximum,}$$

or the sum

$$\Sigma \log f(x_r-x') \text{ a maximum.}$$

Differentiating with respect to x', we obtain the condition for a maximum in the form

$$\Sigma \frac{f'}{f} = 0.$$

Let

$$F(u) = \frac{f'}{f},$$

where u is the error $x_r - x'$.

Then $\Sigma F(u) = 0$ whenever the numbers of positive and negative errors are equal. This is satisfied by making

$$F(u) = \pm k,$$

where the upper or lower sign is to be taken according as u is positive or negative.

Writing

$$\frac{f'}{f} = \pm k$$

and integrating, we find

$$\log f(u) = \pm ku + \text{constant} = k\,|u| + \text{constant},$$

and

$$f(u) = Ce^{k|u|}.$$

Since the probability must decrease as u increases it follows that k is negative, and the form of the error law is

$$f(u) = Ce^{-l^2|u|}.$$

Since any error must lie between $-\infty$ and $+\infty$, we have, as in the case of the Normal Law,

$$1 = \int_{-\infty}^{+\infty} f(u)\,du = 2C\int_0^\infty e^{-l^2 u}\,du = \frac{2C}{l^2},$$

and finally we may write

$$f(u) = \frac{l^2}{2} e^{-l^2|u|}.$$

EXAMPLES.

1. Show that the average value of the error (η of page 30) is $\frac{1}{l^2}$ when the above law holds.

2. Show that the above law yields as the most probable value of the unknown that which makes the arithmetic sum of the errors a minimum.

3. Given that the arithmetic sum of the errors is to be a minimum, show that the median is the most probable value of the unknown.

4. Find the P.E. s of a single observation.

$$[e^{-l^2 s} = \tfrac{1}{2}.]$$

Circular Distributions. The distribution of shots on a target, assuming as in § 9 that the errors along two rectangular axes are independent and follow the same law, is readily seen to be

$$2h^2 r e^{-h^2 r^2} dr \quad \text{or} \quad \frac{d}{dr} e^{-h^2 r^2}.$$

The proportion of shots within a circle of radius r is $(1 - e^{-h^2 r^2})$. Here r is the distance from the centre (mean point of impact). The mean error is

$$2h^2 \int_0^\infty r^2 e^{-h^2 r^2} dr \quad \text{or} \quad \frac{\sqrt{\pi}}{2h}.$$

If r_1 is the circle within which half the shots fall,

$$1 - e^{-h^2 r_1^2} = \tfrac{1}{2}, \qquad r_1 = \frac{\sqrt{\log_e 2}}{h}.$$

Hence $\qquad r_1 = \frac{2\sqrt{\log_e 2}}{\sqrt{\pi}} \times \text{mean error} = 0.94 \times \text{mean error}.$

Mean value of $\qquad r^2 = \int_0^\infty 2h^2 r^3 e^{-h^2 r^2} dr = \frac{1}{h^2} = \frac{\Sigma r^2}{n}.$

$$r_1 = 0.833 \sqrt{\frac{\Sigma r^2}{n}}.$$

REFERENCES.

For a full discussion of a number of attempts at proving the Normal Error Law the reader is referred to Glaisher, *Memoirs of the Royal Astronomical Society*, 1872.

The earlier work on the subject will be found in Todhunter, *History of Probability*.

Gauss's contributions to the subject will be found in: *Theoria Motus Corporum Coelestium*, Lib. II, Sect. III; *Theoria Combinationis Observationum Erroribus Minimis Obnoxiae*, Comment. Soc. Göttingen, Vol. V, 1819–1822; *Supplementum Theoriae Combinationis Observationum, ibid.*, Vol. VI, 1823–1827.

CHAPTER III

17. Measures of Precision.

The area bounded by the axis of x, and the curve

$$y = \frac{h}{\sqrt{\pi}} e^{-h^2 x^2},$$

is unity for all values of h. The greater the value of h, the greater is the central ordinate, and the steeper the curve; i.e. the greater the value of h, the more closely will the observations be clustered about the mean value. If two sets of observations of the same quantity be made, the series for which h is greater will be more closely clustered about the mean value, and may therefore be regarded as a better set of observations than the set for which h is less. Hence h may be regarded as a measure of the precision of the observations, and may be used as a criterion for judging the accuracy with which the observations have been carried out.

In practice it is found more convenient to use certain functions of h, rather than h itself, as measures of precision. Most continental writers use a function which is termed the *mean square error* or M.S.E., a term which is loosely used to denote the square root of the average value of the squares of the errors of observation. It is usually denoted by the Greek letter μ.

Of a set of n observations, the number whose errors will lie between Δ and $\Delta + d\Delta$ will be

$$\frac{nh}{\sqrt{\pi}} e^{-h^2 \Delta^2} d\Delta.$$

The sum of the squares of these errors will be

$$\frac{nh}{\sqrt{\pi}} \Delta^2 e^{-h^2 \Delta^2} d\Delta.$$

If μ be the M.S.E., it follows that the sum of the squares of all the errors from $-\infty$ to $+\infty$

$$= n\mu^2 = \frac{nh}{\sqrt{\pi}} \int_{-\infty}^{+\infty} e^{-h^2\Delta^2} \Delta^2 d\Delta$$

$$= \frac{2nh}{\sqrt{\pi}} \int_0^\infty e^{-h^2\Delta^2} \Delta^2 d\Delta = \frac{n}{2h^2}.$$

$$\therefore \ \mu^2 = \frac{1}{2h^2} \ \text{ and } \ \mu = \frac{1}{\sqrt{2}h}.$$

The error law may thus be written

$$\frac{1}{\mu\sqrt{2\pi}} e^{-\frac{x^2}{2\mu^2}} dx.$$

English and American writers generally use as a measure of precision a quantity r, unhappily named the *probable error* or P.E.* The P.E., r, is of such a magnitude that the error of a single observation is as likely to be within as without the limits $\pm r$. Or, to express it in another way, the odds are even that the error of a single observation shall not be greater in magnitude than r. This condition is expressed mathematically by the equation

$$\frac{1}{2} = \frac{h}{\sqrt{\pi}} \int_{-r}^{+r} e^{-h^2x^2} dx = \frac{2}{\sqrt{\pi}} \int_0^{hr} e^{-x^2} dx.$$

This equation can be solved by the use of tables of the integral $\int_0^x e^{-x^2} dx$, yielding the result

$$hr = 0.47694 = \rho.$$

The quantity ρ is the one mentioned on page 19.
It follows that

$$r = \frac{0.47694}{h} = 0.6745\mu = \tfrac{2}{3}\mu \ \text{(approximately)}.$$

The P.E. has a very simple meaning, in that one-half of the observations in a series should have errors greater than r, and the other half should have errors less than r.

Another measure of precision which is sometimes used is the *average error* η, whose value is the average of the errors of all the

* The *most probable error* is zero, corresponding to the highest point of the error curve.

observations, considered without regard to sign. Its relation to h can be easily derived.

$$n\eta = \frac{nh}{\sqrt{\pi}} \int_0^\infty x\, e^{-h^2 x^2}\, dx + \frac{nh}{\sqrt{\pi}} \int_0^{-\infty} (-x)\, e^{-h^2 x^2}\, dx.$$

$$\therefore \ \eta = \frac{2h}{\sqrt{\pi}} \int_0^\infty x\, e^{-h^2 x^2}\, dx$$

$$= \frac{1}{h\sqrt{\pi}}.$$

It follows that $\qquad \mu = \eta \times \sqrt{\frac{\pi}{2}},$

and $\qquad\qquad r = 0\cdot6745\mu = 0\cdot8453\eta.$

The relations of $\dfrac{1}{h}$, r, μ, and η are here collected in tabular form for convenience of reference.

	$\dfrac{1}{h}$	μ	r	η
$\dfrac{1}{h}$	1·0000	1·4142	2·0966	1·7726
μ	·7071	1·0000	1·4826	1·2533
r	·4769	·6745	1·0000	·8453
η	·5642	·7979	1·1829	1·0000

The following approximate values of some of the above coefficients are occasionally of use :

$$\cdot6745 = \tfrac{2}{3}.$$
$$\cdot8453 = \tfrac{11}{13}.$$
$$\cdot47694 = \tfrac{10}{21}.$$

18. Evaluation of h and r.

The relation between h and the residuals can be obtained as follows. Let n similarly observed values of a quantity be x_1, x_2, x_3, ..., x_n, and let X be the true value of the quantity.

Then the errors of the separate observations are $x_1 - X$, $x_2 - X$, ..., $x_n - X$. The probability of an error lying between x and $x + dx$ is

$$\frac{h}{\sqrt{\pi}} e^{-h^2 x^2} dx.$$

As on page 22 this may be expressed by saying that the probability of an error x is $Che^{-h^2 x^2}$, where C is a constant. Then the *a priori* probability of the occurrence of the system of errors $x_1 - X$, $x_2 - X$, etc. is

$$C^n h^n e^{-h^2 \Sigma (x_r - X)^2},$$

or

$$C^n h^n e^{-h^2 \Sigma v_r^2} e^{-h^2 n (M - X)^2},$$

where

$$M = \frac{\Sigma x}{n} \quad \text{and} \quad v_r = x_r - M.$$

The true value of X is unknown. All that can definitely be said is that it lies between $-\infty$ and $+\infty$. Thus the probability of the given system of measurements $x_1, x_2, ..., x_n$ must be written as the integral

$$C^n h^n e^{-h^2 \Sigma v^2} \int_{-\infty}^{+\infty} e^{-h^2 n (M - X)^2} dX,$$

and this can immediately be reduced to

$$C^n h^{n-1} \sqrt{\frac{\pi}{n}} e^{-h^2 \Sigma v^2}.$$

The value of h must be such as to make this probability a maximum. Taking logarithms and differentiating with respect to h, we find

$$\frac{n-1}{h} - 2h \Sigma v^2 = 0,$$

or

$$h^2 = \frac{n-1}{2\Sigma v^2} = \frac{n-1}{2[vv]}.$$

In writings on the present subject it is customary to denote summation, not by the letter Σ, but by square brackets. The last equation reduces to

$$h = \sqrt{\frac{n-1}{2[vv]}}.$$

It immediately follows that

$$r = 0 \cdot 47694 \frac{1}{h} = 0 \cdot 6745 \sqrt{\frac{[vv]}{n-1}}.$$

19. Comparison of a set of observations with the preceding theory.

Gauss (*Werke*, IV. p. 116) took Bessel's reduction of 470 observations of the right ascensions of Procyon and Altair made by Bradley, and compared the distribution of errors with the theoretical curve obtained by evaluating h by the above formula. He calculated the numbers of observations whose errors should be numerically between $0''\!\cdot\!0$ and $0''\!\cdot\!1$, between $0''\!\cdot\!1$ and $0''\!\cdot\!2$, etc., and compared them with the actual numbers obtained from Bradley's observations. The results are given in the following table.

Errors	Theoretical number	Actual number
$0''\!\cdot\!0$—$0''\!\cdot\!1$	94·8	94
$0''\!\cdot\!1$—$0''\!\cdot\!2$	88·8	88
$0''\!\cdot\!2$—$0''\!\cdot\!3$	78·3	78
$0''\!\cdot\!3$—$0''\!\cdot\!4$	64·1	58
$0''\!\cdot\!4$—$0''\!\cdot\!5$	49·5	51
$0''\!\cdot\!5$—$0''\!\cdot\!6$	35·8	36
$0''\!\cdot\!6$—$0''\!\cdot\!7$	24·2	26
$0''\!\cdot\!7$—$0''\!\cdot\!8$	15·4	14
$0''\!\cdot\!8$—$0''\!\cdot\!9$	9·1	10
$0''\!\cdot\!9$—$1''\!\cdot\!0$	5·0	7
above $1''\!\cdot\!0$	5·0	8

The table shows a remarkable correspondence between the theory and the observational data. There is, however, a slight discrepancy in the number of large errors, the number occurring in practice exceeding the theoretical number. This discrepancy occurs in other series of observations, and some attempt to deal with it will be made in a later chapter.

20. Evaluation of μ.

The quantity μ, the M.S.E. of a system of errors, is connected with h by the relation

$$\mu = \frac{1}{h\sqrt{2}}.$$

It has already been shown that $h = \sqrt{\dfrac{n-1}{2\,[vv]}}$, and so μ should be given by

$$\mu = \sqrt{\frac{[vv]}{n-1}}.$$

This relation is of such importance that it is necessary to consider it in some detail. The residuals v_1, v_2, etc. are the deviations of the observed values from the A.M., and if the A.M. could be definitely regarded as the true value of the unknown, the $(\text{M.S.E.})^2$ ought to be equal to $\dfrac{[vv]}{n}$. The use of the denominator $n-1$ instead of n must be regarded as due to the uncertainty as to the true value of the unknown. The derivation of the formula for μ from a purely algebraic standpoint may help to elucidate the question.

Let $x_1, x_2, \ldots, x_n$ be the n observed values. Now suppose these to be the first n observations in a series of N observations. Let $V_1, V_2, \ldots, V_N$ be the residuals calculated from the mean of the N observations.

Let $$\mu^2 = \frac{[VV]}{N}, \qquad \mu'^2 = \frac{[vv]}{n}.$$

If v_s be the residual of the sth observation within the group of n observations, then

$$v_s = d + V_s \quad \ldots\ldots\ldots\ldots\ldots\ldots(1),$$

where d is difference between the mean of the N observations and the mean of the group of n. The value of d is given by

$$d = \frac{V_1 + V_2 + \ldots + V_n}{n} \ldots\ldots\ldots \ldots\ldots\ldots(2).$$

Squaring equation (2) we find

$$d^2 = \frac{1}{n^2} \sum_1^n V_s^2 + \frac{2}{n^2} \Sigma\, V_s V_t.$$

There will be $\frac{1}{2}n(n-1)$ terms of the form $V_s V_t$, and the equation may be written

$$d^2 = \frac{1}{n^2} \sum_1^n V_s^2 + \frac{n-1}{n} \times \text{mean value of } V_s V_t \quad \ldots\ldots(3).$$

If a large number of different groups of n be selected from among the total of N observations, and equation (3) be formed for each group, we can write down the mean value of each side of equation (3) for all these groups.

The resulting equation is

$$\bar{d}^2 = \frac{\mu^2}{n} + \frac{n-1}{n} \times \text{mean value of } V_s V_t \ldots\ldots\ldots\ldots(4),$$

where $\bar{d}^2$ is the mean value of d^2.

But
$$0 = (V_1 + V_2 + \ldots + V_N)^2$$
$$= [VV] + N(N-1) \times \text{mean value of } V_s V_t.$$

$\therefore$ mean value of $V_s V_t = -\dfrac{[VV]}{N(N-1)} = -\dfrac{\mu^2}{N-1}$ ...(4').

Substituting this result in equation (4) we find

$$\bar{d}^2 = \frac{\mu^2}{n} - \frac{n-1}{n}\frac{\mu^2}{N-1} = \frac{\mu^2(N-n)}{n(N-1)} \quad \ldots\ldots\ldots\ldots(5).$$

Now
$$V_s^2 = (d + v_s)^2 = d^2 + 2dv_s + v_s^2.$$

There will be n such equations, and if we write these n equations down and take the mean value of both sides, we find

$$\text{mean value of } V_s^2 = d^2 + \mu'^2 \quad \ldots\ldots\ldots\ldots(6).$$

If we form equation (6) for all the possible sets of n observations, and then take the mean of all these equations, we find

$$\mu^2 = \bar{d}^2 + \mu'^2,$$

or, substituting for $\bar{d}^2$ from equation (5),

$$\mu^2 = \frac{\mu^2}{n}\frac{N-n}{N-1} + \mu'^2,$$

$$\mu'^2 = \mu^2\frac{n-1}{n}\frac{N}{N-1} \quad \ldots\ldots\ldots\ldots(7).$$

If we have a very large number N of observations, and we take samples of n at a time, from among the number N, the average value of the (mean-square-residual)^2 for all the possible samples is connected with the mean square residual of all the N observations (μ), by equation (7).

If N be regarded as becoming an indefinitely large number, the mean of the N observed values may be taken as the true value of the unknown, and μ is then the M.S.E. of a single observation. Equation (7) then yields

$$\mu^2 = \frac{n}{n-1}\mu'^2 = \frac{n}{n-1}\frac{[vv]}{n} = \frac{[vv]}{n-1}.$$

Whence
$$\mu = \sqrt{\frac{[vv]}{n-1}}.$$

The formula $(n-1)\mu^2 = [vv]$ may be interpreted as follows. If we take an infinite series of observations of any quantity, and select a large number of samples of n observations, the sum of the squares of the n residuals calculated from the mean of the n observations will have the mean value $(n-1)\mu^2$. The factor $n-1$ is due to the fact that the mean value of the observations in any one sample is liable to differ from the true value of the unknown.

21. Probable error and mean square error of the arithmetic mean.

Returning to equation (5) above, $\bar{d}$ is the M. S. E. of the mean of n observations. If we take N to be an infinitely large number, μ becomes the M. S. E. of a single observation. Equation (5) may then be written

$$\bar{d}^2 = \frac{\mu^2}{n} \text{ or } \bar{d} = \frac{\mu}{\sqrt{n}},$$

or   M. S. E. of A. M. $= \dfrac{\text{M.S.E. of a single observation}}{\sqrt{n}}$.

Since the relation $r = 0\cdot6745\mu$ is always true, we obtain immediately the result

$$\text{P.E. of A. M.} = \frac{\text{P. E. of a single observation}}{\sqrt{n}} = \frac{r}{\sqrt{n}}$$

$$= 0\cdot6745 \sqrt{\frac{[vv]}{n(n-1)}}.$$

22. Probable error of a linear function of a number of independent quantities whose probable errors are known*.

In the first case we shall consider a linear function of two independent quantities m_1, m_2, whose mean square errors are μ_1, μ_2. Let the linear function be

$$F = a_1 m_1 + a_2 m_2,$$

where a_1 and a_2 are constants.

If an error x be made in determining m_1, and an error y in determining m_2, the corresponding error dF in F is given by

$$dF = a_1 x + a_2 y.$$

Squaring this equation, we obtain

$$dF^2 = a_1^2 x^2 + a_2^2 y^2 + 2a_1 a_2 xy.$$

---

* This proof assumes that the error law of a linear function of a number of independent quantities is of the same form as the error law of each of these quantities. *Vide* § 15 for proof of this assumption.

This equation holds for all values of x and y; and therefore the mean value of the L. H. S. is equal to the mean value of the R. H. S. But if μ, μ_1, μ_2 be the M. S. E.'s of F, m_1, and m_2, respectively, the equation leads to

$$\mu^2 = a_1^2\mu_1^2 + a_2^2\mu_2^2 + 2a_1a_2 \times \text{mean value of } xy.$$

Since m_1 and m_2 are independent, the errors x and y are also independent. With a given value of x, positive and negative values of y are equally likely to be associated; and so the mean value of the product xy is zero.

$$\therefore \ \mu^2 = a_1^2\mu_1^2 + a_2^2\mu_2^2.$$

This result may be extended to apply to any number of variables, the proof being the same as that given above for two variables. In general, if

$$F = a_1m_1 + a_2m_2 + \ldots + a_nm_n,$$

where a_1, a_2, ..., a_n are constant, then

$$\mu^2 = a_1^2\mu_1^2 + a_2^2\mu_2^2 + \ldots + a_n^2\mu_n^2,$$

or, if r, r_1, r_2, ..., r_n be the P.E.'s of F, m_1, m_2, ..., m_n,

$$r^2 = a_1^2r_1^2 + a_2^2r_2^2 + \ldots + a_n^2r_n^2.$$

Probable error of the Arithmetic Mean.

The P. E. of the A. M. can be immediately derived from the above. For if x_1, x_2, ..., x_n be n independently observed values, each with P. E. r, and if r' be the P. E. of the A. M. $\bar{x}$, then

$$\bar{x} = \frac{1}{n}(x_1 + x_2 + \ldots + x_n),$$

and

$$r'^2 = \frac{1}{n^2}(r^2 + r^2 + \text{etc. to } n \text{ terms}) = \frac{r^2}{n}.$$

The P. E. of the A. M. is therefore $\dfrac{r}{\sqrt{n}}$.

If μ be the M. S. E. of a single observation, the M.S.E. of the A. M. is $\dfrac{\mu}{\sqrt{n}}$.

Exercise. Derive the above results from the formulae given at the end of § 15.

23. Peters' Formula for r.

Let x_1, x_2, ..., x_n be n observed values of a quantity whose true value is x. Then the true errors ϵ_1, ϵ_2, ..., ϵ_n are given by the equations

$$x_1 = x - \epsilon_1, \quad x_2 = x - \epsilon_2, \text{ etc.}$$

The A. M. of the observed values is

$$\frac{1}{n}(x_1 + x_2 + \ldots + x_n),$$

or

$$x - \frac{1}{n}(\epsilon_1 + \epsilon_2 + \ldots + \epsilon_n).$$

If v_1, v_2, etc. be the deviations of x_1, x_2, etc. from the mean,

$$v_1 = \frac{1}{n}\{(n-1)\epsilon_1 - \epsilon_2 - \epsilon_3 - \ldots - \epsilon_n\},$$

$$v_2 = \frac{1}{n}\{-\epsilon_1 + (n-1)\epsilon_2 - \epsilon_3 - \ldots - \epsilon_n\}, \text{ etc.}$$

All the n observations are supposed to be liable to the same errors, or are supposed to follow the same error law. And since the residuals v_1, v_2, etc. are linear functions of ϵ_1, ϵ_2, etc., it follows from § 15 that the residuals are subject to a similar error law, the parameter h being the same for all the residuals, since the sum of the squares of the coefficients on the R. H. S. is the same for all v's.

Let the P. E. of each ϵ be r, and let the P. E. of any residual v be r'. Then it follows from § 22 that

$$r'^2 = \frac{r^2}{n^2}\{(n-1)^2 + (n-1)\} = r^2\frac{n-1}{n} \quad \ldots\ldots\ldots\ldots(1).$$

Hence the true P. E. r of the observations is $r'\sqrt{\dfrac{n}{n-1}}$, where r' is the P. E. derived from the residuals.

If $v]$ be the sum of the residuals, taken without regard to sign, then

$$r' = 0{\cdot}8453\,\frac{v]}{n}. \qquad \text{(See page 31.)}$$

$$\therefore \quad r = 0{\cdot}8453\,\frac{v]}{\sqrt{n(n-1)}} \quad \ldots\ldots\ldots\ldots\ldots(2).$$

Equation (2) is known as Peters' formula for the probable error.

It should be noted that equation (1) above leads also to the ordinary formula for the P. E. derived from the squares of the residuals. The P. E. of a residual

$$= r' = 0.6745 \sqrt{\frac{[vv]}{n}},$$

$$\therefore \ r = r' \sqrt{\frac{n}{n-1}} = 0.6745 \sqrt{\frac{[vv]}{n-1}} \ \ldots \ldots \ldots (3).$$

We thus have two equations for the evaluation of the P. E. of a single observation from the residuals.

24. Some Examples of the Adjustment of Observations of one unknown.

Example 1.   *The heat of evaporation of water*.

Each of the 20 values given in the first column of the table below is an independent determination of the heat of evaporation of water. It is required to find the adjusted value, and its P. E. The A.M. of all the observations is adopted as the most probable value. The residuals obtained by subtracting the A.M. from each determination are written in the second column, the positive and negative residuals being separated into sub-columns. The $\frac{1}{4}$ squares of the residuals are written in the last column.

Observed value	Residual +	Residual −	$\frac{1}{4}$ square of residual
542·98	1·056		·2788
1·23		·694	·1204
0·64		1·284	·4122
2·03	·106		·0028
2·32	·396		·0392
1·48		·444	·0493
2·37	·446		·0497
2·15	·226		·0128
1·36		·564	·0795
1·34		·584	·0853
2·91	·986		·2430
2·68	·756		·1429
3·08	1·156		·3341
2·12	·196		·0096
1·82		·104	·0027
0·96		·964	·2323
1·66		·264	·0174
1·73		·194	·0094
1·79		·134	·0045
1·83		·094	·0022
Mean 541·924	+5·324	−5·324	2·1281

* A. W. Smith, *Physical Review*, September, 1911.

The correspondence of the sums of positive and negative residuals affords a check on the value of the A.M.

$$[vv] = 2 \cdot 1281 \times 4 = 8 \cdot 5124.$$

Using the formulae of pages 32 and 36, we find

$$\text{P.E. of a single observation} = \cdot 6745 \sqrt{\frac{8 \cdot 5124}{19}} = \cdot 451,$$

$$\text{P.E. of A.M.} = \frac{\cdot 451}{\sqrt{20}} = \cdot 101.$$

Using Peters' formula we find

$$\text{P.E. of a single observation} = \cdot 8453 \times \frac{10 \cdot 648}{\sqrt{20 \times 19}} = \cdot 462,$$

$$\text{P.E. of A.M.} = \frac{\cdot 462}{\sqrt{20}} = \cdot 103.$$

The two formulae for P.E. yield almost identical results. The final value of the unknown, together with its P.E., can be represented as

$$541 \cdot 924 \pm \cdot 102.$$

The residuals of column 2 are shown in figure 4. The whole range of variation is divided up into intervals of ·4, the middle interval extending

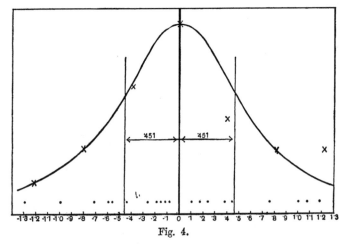

Fig. 4.

from − ·2 to + ·2. In the diagram a dot is put in to represent, as accurately as possible, the value of each residual. An ordinate is erected at the middle of each interval, of a length proportional to the number of observations falling within that interval. The top of each ordinate is represented by a ×. A smooth symmetrical curve is drawn to fit the tops as closely as possible. Although two of the points representing positive residuals are at some distance from the curve, it is seen that the general form of the curve is in fair agreement with the Gauss error curve shown in figure 3. The discrepancy is possibly due to the smallness of the number of determinations used in the reduction.

In the diagram two ordinates are erected at distances $\pm\,\cdot451$ from the origin. According to the least square theory, 10 observed values should lie between these ordinates, and 10 outside these limits. It is seen that 11 observed values lie within the given limits, and 9 without; a result in sufficiently good agreement with the theory, in view of the relative smallness of the number of determinations.

Example 2. The atomic weight of bromine*.

The ratio of the weights of Bromine and Hydrogen which combine to form Hydrobromic acid was determined experimentally. The results of 10 independent determinations are given in the first column of the table below.

Observed value	Residual		¼ square of residual
	+	−	
79·2863		·0204	·00010404
·3055		·0012	36
·3064		·0003	2
·3197	·0130		4225
·3114	·0047		552
·3150	·0083		1722
·3063		·0004	4
·3141	·0074		1369
·2915		·0152	5776
·3108	·0041		420
Mean 79·3067	·0375	·0375	·00024510

$$[vv] = \cdot0009804, \quad v] = \cdot075,$$

$$\text{P. E. of a single determination} = \sqrt{\frac{\cdot0009804}{9}} \times \cdot6745 = \cdot0070,$$

$$\text{P. E. of the A.M.} = \frac{\cdot0070}{\sqrt{10}} = \cdot0022,$$

$$\text{P.E. of A.M. from Peters' formula} = \frac{1}{\sqrt{10}} \times \cdot8453 \times \frac{\cdot075}{\sqrt{90}} = \cdot0021.$$

The adjusted value of the ratio is thus $79\cdot3067 \pm \cdot002$.

The atomic weight of Hydrogen $= 1\cdot00779$.

Therefore the atomic weight of Bromine

$$= 1\cdot00779\,(79\cdot3067 \pm \cdot002) = 79\cdot924 \pm \cdot002.$$

* Weber, *Bulletin of Bureau of Standards*, Vol. IX, p. 131.

Example 3. *The percentage of dry matter in mangel roots**.

The percentage of dry matter was estimated for each of 160 roots of a strain of Golden Globe mangel. The results varied between 10·7 °/₀ and 19·7 °/₀, and the A.M. of all the results gave 14·5 °/₀. It was necessary to consider whether it was justifiable to take the A.M. of such widely differing results, and if so, what was its precision. In order to find the answer to this question we must first consider what are the different causes which tend to produce differences in content of dry matter in individual roots; and whether these causes satisfy the assumptions made on page 5 as to the nature of accidental errors.

All the roots taken were of the same strain, grown side by side, and sampled and analysed in the same manner. Thus the possibilities of variation in individual roots were reduced to a minimum. But there still remained certain possible causes of variation. Since mangels are easily cross-fertilised, a commercial strain will not be a pure one, and so varying parentage may be a possible cause of variation in constitution. The slight differences in soil will vary the food supply; the hoeing will not be absolutely regular; the distribution of manure will not be perfectly uniform, and slight errors may enter into the analyses of the roots. Each of these separate causes is equally likely to make the percentage of dry matter in an individual root higher or lower than the average value. In the majority of cases, some of these causes will tend to raise, and some to lower the result. It will only happen in relatively few cases that all the causes of variation act in the same direction, yielding a result differing considerably from the mean value. A curve showing the frequencies of different percentages of dry matter should thus have its maximum at the mean value of the percentage, and should be symmetrical about the mean value. It might, in fact, be expected to yield a curve of the same general form as the error curve shown in figure 3.

The actual distribution of frequencies is shown in figure 5. In this diagram, the percentage of dry matter is represented along the horizontal axis, and the number of roots along the vertical axis. For each root a dot is placed in the diagram above the corresponding percentage. The whole range of variation is divided into intervals, from 10—11 °/₀, 11—12 °/₀, etc., and an ordinate is erected at the middle point of each section, of a length proportional to the number of dots in that section. The tops of these ordinates are represented in the diagram by crosses. It is clear from inspection that it would be possible to draw a smooth symmetrical curve passing very near to the tops of the ordinates. The form of such a curve would be in good agreement with that of the ideal error curve shown in figure 3. To test this more closely, the Gauss error curve was drawn for comparison with the actual distribution. This curve is the smooth curve

* Wood and Stratton, "The Interpretation of Experimental Results," *Journal of Agricultural Science*, Vol. III, part 4.

shown in figure 5. The method of construction of this curve will be explained later on, but meanwhile we may note that the crosses representing the actual distribution all lie very near to the curve.

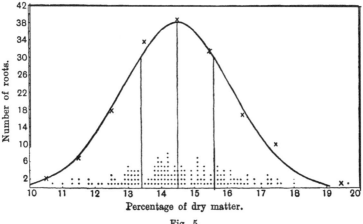

Fig. 5.

The P.E. of a single determination is given as $1\cdot1 \,\%$.*. Thus the P.E. of the A.M. is $\dfrac{1\cdot1}{\sqrt{160}} = 0\cdot09$. In the diagram, ordinates are drawn on each side of the mean, distant $1\cdot1$ from it. According to the theoretical discussion, 80 of the observed values should lie within, and 80 outside these limits. A count of the dots shows that 81 observations are within $1\cdot1 \,\%$ of the mean, and 79 outside these limits—a result in excellent agreement with the demands of the theory.

The equation of the normal error curve is

$$y = \frac{h}{\sqrt{\pi}} e^{-h^2 x^2},$$

where
$$h = \frac{\cdot47694}{r} = \frac{\cdot47694}{1\cdot1} = \cdot43.$$

A sufficient number of ordinates to enable us to draw the curve can easily be evaluated by means of a table of logarithms.

In the preceding example, the symmetry of the distribution of frequencies about the A.M. appears to justify the assumption that the different causes of variation in content of individual roots were equally likely to raise as to lower the result. In general, when some of the causes of variation are dissymmetrical in their action, tending to give high results oftener than low results, or vice versa, the frequency curve is unsymmetrical, even when the

---

* *Loc. cit.*

number of observations is sufficiently large to give the laws of chance fair play. If the number of observations be large, and the curve unsymmetrical, it is generally safe to assume that there is at work some cause which tends to act always in one direction, giving an abnormal number of high or of low results. Such a case is discussed in Example 4 below.

Example 4. The average weight of a number of mangel roots*.

The figure shows the distribution of the weights of 196 roots. In this case no smooth curve has been drawn, but the tops of successive ordinates have been joined by straight lines. The resulting curve is clearly dissymmetrical, showing that either an abnormally large number of large roots, or an abnormally small number of small roots, was taken. The latter alternative probably affords the true explanation, as the weak plants which would produce small roots would be destroyed in the process of hoeing and singling;

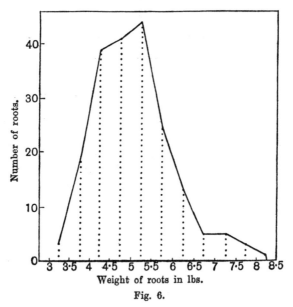

Fig. 6.

and it is possible that the very small roots would be unconsciously passed over in sampling. The true curve of frequency should therefore extend further in the direction of small weights. The effect of the absence of small roots is to give an apparent mean weight greater than the true value. The method of least squares is not strictly applicable to such a distribution as is here shown, as the theory demands a distribution of frequencies which shall be symmetrical about the mean.

* Wood and Stratton, *loc. cit.*

Example 5. *Magnitude-interval of a parallel-wire grating**.

A parallel-wire grating consists of a series of equidistant parallel wires fitted on a frame. When this is set over the object glass of a telescope, it acts as a diffraction-grating, so that each star, instead of producing a single dot upon the photographic plate, produces a central bright dot, with a series of dots of decreasing brightness on each side of it. The dots are in reality small spectra, but on account of the smallness of the dispersion they are sensibly round. The proportion of the light of a star which is deviated into an image of a given order is constant, depending only on the form of the grating. There is accordingly a definite magnitude-interval between a central image and the auxiliary image on each side of it. There will be this same magnitude-interval between a given star and another star whose central image is equal in size and greyness to the first diffraction image of the first star. Thus when the magnitude-interval between consecutive diffraction images is known, the magnitudes of all the stars on the plate can be determined, provided there are on the plate some stars whose magnitudes are known. The adjoining table shows a series of determinations of the magnitude-interval of a grating, based on measurements of a number of stars upon plates with different exposures.

No. of plate...	4996	5014	5061	4922	4923	4924	4997	5059	5070	5023
Magnitude of star										
8·89	2·68	2·66	3·09	2·66	2·73	2·31	2·65	2·67	2·94	—
10·42	2·67	2·90	2·62	2·67	2·75	2·86	2·98	2·60	2·68	2·57
10·54	2·74	2·86	2·84	2·68	2·66	—	2·67	2·46	2·60	2·75
10·62	2·56	2·54	2·64	2·78	2·62	2·69	2·72	2·65	2·59	—
10·64	2·49	2·58	2·77	2·70	2·70	2·77	2·66	2·72	2·50	—
10·64	2·56	2·69	2·62	2·70	2·70	2·73	2·66	2·61	2·46	—
10·66	—	—	—	2·76	2·63	2·64	—	—	—	—
10·69	—	—	2·72	2·74	2·66	2·94	—	—	2·82	—
10·90	2·78	2·73	2·91	2·64	3·03	—	—	—	—	—
10·94	2·82	—	2·85	—	—	—	—	—	—	—
10·95	2·79	2·95	2·77	2·60	2·59	2·99	—	—	—	—
11·01	—	2·69	2·89	—	—	—	—	—	—	—
11·03	2·76	2·79	2·62	2·55	2·97	—	—	—	—	—
11·12	2·65	—	2·80	2·41	2·87	2·86	—	—	—	—
11·26	2·67	—	2·77	2·32	2·87	—	—	—	—	—
11·47	3·00	—	—	—	—	—	—	—	—	—
11·57	2·68	—	—	—	—	—	—	—	—	—
11·97	2·63	—	—	—	—	—	—	—	—	—
12·12	2·54	—	—	—	—	—	—	—	—	—
12·25	2·83	—	—	—	—	—	—	—	—	—
12·32	2·79	—	—	—	—	—	—	—	—	—

* Chapman and Melotte, *Monthly Notices, R.A.S.*, November, 1913.

The mean of 98 determinations $= 2^{m}\cdot71$.

The P. E. of a single estimate calculated from the squares of the residuals
$$= \pm 0^{m}\cdot097.$$

The P. E. of the mean value is $\dfrac{\cdot097}{\sqrt{98}} = \cdot0097$.

The final result may thus be written
$$2^{m}\cdot71 \pm \cdot0097.$$

The P. E. of a single estimate calculated from the sum of the residuals is $\cdot094$.

The frequency curve is shown in figure 7. It is seen that the A.M. does not coincide with the maximum ordinate. The latter occurs at about $2^{m}\cdot67$, whereas the A. M. is $2^{m}\cdot71$. The number of measurements represented by the

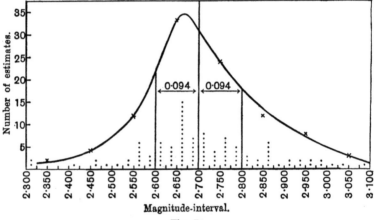

Fig. 7.

curve is 98, and this appears to be a sufficiently large number to yield an accurate representation of the true nature of the curve of errors. We are forced to conclude that there is present some cause of dissymmetry.

The P.E. of a single observation being $0^{m}\cdot097$, two ordinates are drawn, one on each side of the A. M., and at a distance of $0^{m}\cdot097$ from it. It is found that 57 measurements lie between these limits, and 41 outside them. Theory would require 49 observations to lie within the limits $\pm 0^{m}\cdot097$, and 49 outside these limits. It is thus no longer strictly possible to attach the original meaning to the P.E. deduced from the residuals on the assumption that the mean is the true value of the unknown.

When the curve of frequencies is of the form of the curve in figure 7, indicating a genuine dissymmetry in the distribution of frequencies, it would perhaps be better to adopt the abscissa corresponding to the maximum frequency, as the most plausible value of the unknown. The value of the

P. E. deduced from the formula may be taken to indicate roughly whether the observations are closely clustered about the mean, or are spread over a considerable range. But this P. E. cannot be regarded as having the meaning originally attached to the P. E., since the curve representing the frequency distribution is not of the form of the Gauss error curve represented in figure 3.

Example 6.   Glume-length of wheat.

Figure 8 shows the results of measuring the length of the glumes of 595 individual wheat plants. The curve shows three well-defined maxima, and this fact in itself would arouse suspicion as to the purity of strain of the wheat measured. The plants measured were, as a matter of fact, the second generation from a cross between Rivet wheat (with an average glume-

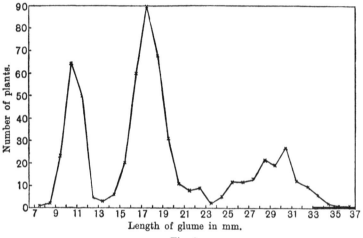

Length of glume in mm.

Fig. 8.

length of 9 mm.) and Polish wheat (with an average glume-length of 28 mm.). The curve shows that the plants examined are divided into three well-defined groups, one resembling the short-glumed Rivet parent, one resembling the long-glumed Polish parent, and the third intermediate between these two with an average glume-length of 17 mm.

The application of the ordinary least-square method to such material as is here represented is quite meaningless. This particular example illustrates very clearly the utility of commencing the discussion of a number of observations by drawing the curve of frequencies. Here the non-homogeneous nature of the material is immediately shown by the curve: while in cases like those considered in Examples 4 and 5 above, the curve of frequencies shows the presence of a systematic error.

EXAMPLES.

In the following Examples calculate the P.E. from both Gauss's and Peters' formula.

1. Evaluate the P.E. of a single determination, and of the mean value, of the variable tabulated on page 6.

2. The following table gives 12 determinations of the azimuth of Allen from Sears, Texas. (U.S. C. and G. Survey. Publications, No. 14, p. 149.)

$$98° \quad 6' \quad 41''\!\cdot\!5$$
$$42''\!\cdot\!8$$
$$43''\!\cdot\!4$$
$$43''\!\cdot\!1$$
$$39''\!\cdot\!7$$
$$42''\!\cdot\!7$$
$$41''\!\cdot\!6$$
$$43''\!\cdot\!3$$
$$40''\!\cdot\!0$$
$$45''\!\cdot\!0$$
$$43''\!\cdot\!3$$
$$40''\!\cdot\!7$$

Find the mean value, the P.E. of a single determination, and the P.E. of the mean.

3. From the following 15 independent determinations of the coefficient of expansion of dry air (Rudberg, *Poggendorff's Annalen*, 41, p. 271), find the A.M. and its P.E.:

$3\cdot643 \times 10^{-3}$	$3\cdot636 \times 10^{-3}$	$3\cdot646 \times 10^{-3}$
54	51	3·662
44	43	3·840
50	43	3·902
53	45	3·652

25. Probable error of any function of a number of independent quantities whose probable errors are known.

Let $m_1, m_2, \ldots, m_n$ be n independent quantities, whose P.E.'s are $r_1, r_2, \ldots, r_n$, and let

$$F = f(m_1, m_2, \ldots, m_n)$$

be the function of m_1, m_2, etc. whose P.E. is required.

If dF be the error in the value of F produced by errors dm_1, dm_2, etc. in the values of m_1, m_2, etc., then

$$F + dF = f(m_1 + dm_1, m_2 + dm_2, \ldots, m_n + dm_n)$$
$$= f + \frac{\partial f}{\partial m_1} dm_1 + \frac{\partial f}{\partial m_2} dm_2 + \ldots + \frac{\partial f}{\partial m_n} dm_n.$$

The error in F is thus given by

$$dF = \frac{\partial f}{\partial m_1} dm_1 + \frac{\partial f}{\partial m_2} dm_2 + \ldots + \frac{\partial f}{\partial m_n} dm_n,$$
$$= a_1 dm_1 + a_2 dm_2 + \ldots + a_n dm_n,$$

where $\quad\quad a_1 = \dfrac{\partial f}{\partial m_1}, \quad a_2 = \dfrac{\partial f}{\partial m_2}$, etc.

From this point onward the problem is reduced to that of finding the P.E. of a linear function of n independent variables.

The result may be written

$$r^2 = a_1{}^2 r_1{}^2 + a_2{}^2 r_2{}^2 + \ldots, \text{ etc.,}$$
or $\quad\quad \mu^2 = a_1{}^2 \mu_1{}^2 + a_2{}^2 \mu_2{}^2 + \ldots, \text{ etc.,}$

where $\quad\quad\quad\quad a_s = \dfrac{\partial f}{\partial m_s}.$

The above formulae sometimes break down in practice, when the errors are not small enough to justify the neglect of the square and higher terms in the Taylor expansion.

EXAMPLES.

1. If x and y be the sides of a rectangle, and r_1, r_2 their P.E.'s, find the P.E. r of the area of the rectangle.

Let $\quad\quad\quad\quad z = xy.$
Then $\quad\quad\quad\quad dz = y\,dx + x\,dy.$

Applying the formula derived above, we find

$$r^2 = y^2 r_1{}^2 + x^2 r_2{}^2.$$

2. The edge of a cube is of length a, and its P.E. is r. Find the P.E. of the volume of the cube. *Ans.* $3a^2 r.$

3. The P.E.'s of the three edges of a cube are r_1, r_2, r_3. Find the P.E. of the volume. *Ans.* $a^2 (r_1{}^2 + r_2{}^2 + r_3{}^2)^{\frac{1}{2}}.$

4. Given that the P.E. of x is r, find the P.E.'s of e^{-x}, $\log x$, $\cos x$, and $(4 + x)^3$.

5. If the P.E. of a single reading of the graduated circle on a meridian circle be r, what is the P.E. of a result based on the readings of n microscopes placed at n different points of the circle ? *Ans.* $\dfrac{r}{\sqrt{n}}.$

6. Given the following telegraphic longitude determinations, find the longitude of Moscow east of Greenwich, and estimate its P.E.

Potsdam—Greenwich	...	...	0h	52m	16s·051 ± 0s·003.
Pulkowa—Potsdam	...	...	1	9	2·491 ± 0·003.
Moscow—Pulkowa	...	...	0	28	58·450 ± 0·010.

7. Given the sides a, b, and the angle C of a triangle, and their P.E.'s, find the P.E.'s of the side c, and of the area of the triangle.

(i) $c^2 = a^2 + b^2 - 2ab \cos C,$

$c\,dc = (a - b \cos C)\,da + (b - a \cos C)\,db + ab \sin C\,dC$

$\qquad = c \cos B\,da + c \cos A\,db + ab \sin C\,dC;$

$\therefore \quad dc = \cos B\,da + \cos A\,db + a \sin B\,dC$

Applying the formula derived above, we find

$$r_c^2 = r_a^2 \cos^2 B + r_b^2 \cos^2 A + r_C^2 a^2 \sin^2 B \sin^2 1'',$$

where r_C is measured in seconds of arc.

(ii) $\Delta = \tfrac{1}{2}ab \sin C,$

$$\frac{d\Delta}{\Delta} = \frac{da}{a} + \frac{db}{b} + \cot C\,dC,$$

$$\frac{r_\Delta^2}{\Delta^2} = \frac{r_a^2}{a^2} + \frac{r_b^2}{b^2} + \cot^2 C \sin^2 1'' r_C^2.$$

8. The coefficient of expansion of a rod is determined by measuring its length at two different temperatures. If the length be l_1 at a temperature t_1, and l_2 at a higher temperature t_2, the coefficient is given by

$$a = \frac{l_2 - l_1}{l_1(t_2 - t_1)}.$$

If r_t be the P.E. of a temperature-reading, and r_l the P.E. or a length-determination, find the P.E. of a.

The P.E. of $l_2 - l_1$ is $\sqrt{2}\,r_l$, and P.E. of $t_2 - t_1$, $\sqrt{2}\,r_t$. Taking logarithms of both sides of the equation for a, and differentiating, we obtain

$$\frac{da}{a} = \frac{d(l_2 - l_1)}{l_2 - l_1} - \frac{dl_1}{l_1} + \frac{d(t_2 - t_1)}{t_2 - t_1},$$

whence we find

$$\frac{r_a^2}{a^2} = \frac{2r_l^2}{(l_2 - l_1)^2} + \frac{r_l^2}{l_1^2} + \frac{2r_t^2}{(t_2 - t_1)^2}$$

$$= r_l^2\left(\frac{1}{l_1^2} + \frac{2}{(l_2 - l_1)^2}\right) + \frac{2r_t^2}{(t_2 - t_1)^2}.$$

26. Errors due to separable causes.

When the accidental errors which enter into a measurement can be divided into a number of separate parts, each of which is independent of the others, and has a known M.S.E. or P.E., the M.S.E. or P.E. of the whole error can be subdivided into the same number of parts. For if an error ϵ be composed of three separate parts ϵ_1, ϵ_2, ϵ_3, all independent of one another, we have

$$\epsilon = \epsilon_1 + \epsilon_2 + \epsilon_3.$$

Then it follows as in § 22 that

$$\mu^2 = \mu_1{}^2 + \mu_2{}^2 + \mu_3{}^2,$$

and
$$r^2 = r_1{}^2 + r_2{}^2 + r_3{}^2.$$

EXAMPLE.

A base line is measured by successive end to end placings of a rod, and is found to be 100 times the length of the rod. If a be the P.E. of the assumed length of the rod, due to uncertainty of the assumed temperature, and b be the P.E. of the end to end placings of the rod, and c the P.E. of the settings at the extremities of the base line, find the P.E. of the assumed length of the base.

(1) If the error in the assumed length of the rod be x, the resulting error in the length of the base will be $100x$. The P.E. of the length of the base, having regard to this class of error only, will be $100a$.

(2) There will be 100 settings of the rod, and the P.E. which enters at each of the 99 intermediate points will be b. The P.E. of the length of the base, if this be the only kind of error, will be $\sqrt{99}\,b$.

(3) The P.E. of the length of the base, due to the end readings, will be $\sqrt{2}\,c$.

Thus, if r be the total P.E. of the length of the base,

$$r^2 = 100^2 a^2 + 99 b^2 + 2c^2.$$

27. Total Probable Error when a Systematic Error is present.

The case where the total error in an observation is partly due to accidental causes, and partly due to systematic causes, follows naturally from the discussion of the last paragraph.

Let r = P.E. of an observation, when the accidental errors only are considered,

r_0 = the mean error arising from the systematic causes,

r' = total P.E. of such an observation.

Then
$$r'^2 = r^2 + r_0{}^2,$$
as in § 22.

If the observation be repeated n times, and the mean taken the final P.E. r'' is given by

$$r''^2 = \frac{r^2}{n} + r_0{}^2.$$

The systematic error is in no way affected by the repetition of the observation, while the P.E. due to the accidental causes is decreased to $\dfrac{r}{\sqrt{n}}$.

The correct interpretation of the last equation is of vital importance in all applications of our present subject. If only accidental errors are present, and if r be the P.E. of a single observation, then the P.E. of the mean of n observations is $\dfrac{r}{\sqrt{n}}$. This expression would lead one to expect that it would be possible to obtain a result free from error simply by increasing the number of observations. In practice, however, it is impossible to eliminate all traces of systematic errors. If the systematic error be represented by r_0, we have seen above that the total P.E. of the mean of n observations is $\sqrt{\dfrac{r^2}{n} + r_0^2}$. It may happen that when n is relatively small r_0^2 is negligible in comparison with $\dfrac{r^2}{n}$; but so long as r_0 is finite, however small it may be, by increasing the value of n we shall eventually reach a stage where $\dfrac{r^2}{n}$ becomes negligible in comparison with r_0^2. When such a stage has been reached, no further improvement in the precision of the mean value can be attained by increasing the number of observations. Or, to put this in other words, as the number of observations is made greater and greater, a stage is eventually reached beyond which the error of the result is fixed by the constant or systematic errors, and not by the probable accidental error. A further increase in the number of observations produces no corresponding increase in the accuracy of the result, unless the conditions of observation can be varied from time to time, so as to vary the magnitude and sign of the systematic errors, thus causing them to appear effectively as accidental errors.

The P.E. computed from the residuals is independent of the presence of a constant error. For if a constant quantity be added to all the observed values in a series, the mean value is altered by the same amount, and the residuals (and the P.E.) remain unchanged. The probable error is a measure, *not* of the deviations of the observed values from the true value, but of their deviations from the mean of an infinite number of observations.

28. The Correction of Statistics for the Effects of a known Probable Error of Observation.

When a table is constructed to exhibit the distribution of frequencies of successive values of a certain measured property whose P.E. is known, it is necessary to consider the effect of the probable error upon the table. For simplicity's sake we shall consider in detail the problem discussed by Eddington* in his treatment of this problem.

Suppose we have a table giving the results of the counts of stars between given limits of magnitude, and suppose the P.E. of a magnitude determination is known. Let this P.E. be $0.477 \dfrac{1}{h}$, so that the probability of an error x is $Ce^{-h^2 x^2}$.

Let $u(m)\, dm$ = observed number of stars between magnitudes

$$m \text{ and } m + dm,$$

$v(m)\, dm$ = true number.

Of the stars whose true magnitude is between $(m + x)$ and $(m + x + dx)$ the proportion $\dfrac{h}{\sqrt{\pi}} e^{-h^2 x^2} dx$ will have an error between $-x$ and $-(x + dx)$, and will be observed as of magnitude m. Thus we have

$$u(m) = \frac{h}{\sqrt{\pi}} \int_{-\infty}^{+\infty} v(m + x)\, e^{-h^2 x^2}\, dx.$$

By the symbolic form of Taylor's theorem

$$v(m + x) = e^{x \frac{d}{dm}} v(m),$$

and therefore

$$u(m) = \frac{h}{\sqrt{\pi}} \int_{-\infty}^{+\infty} e^{x \frac{d}{dm} - x^2 h^2} v(m)\, dx.$$

Treating this integral as a special case of

$$\int_{-\infty}^{+\infty} e^{-a_1 x - a_2 x^2}\, dx = \sqrt{\frac{\pi}{a_2}} \exp \frac{a_1^2}{4 a_2},$$

---

* *Monthly Notices, R.A.S.*, Vol. LXXIII, pp. 359, 360, from which the whole of this discussion has been taken.

we find

$$u(m) = \exp\left(\frac{1}{4h^2}\frac{d^2}{dm^2}\right)v(m)$$

and

$$v(m) = \exp-\left(\frac{1}{4h^2}\frac{d^2}{dm^2}\right)u(m)$$

$$= u(m) - \frac{1}{4h^2}u''m + \frac{1}{2!}\left(\frac{1}{4h^2}\right)^2 u^{iv}(m) - \text{etc.}$$

When the P.E. is small, it is sufficient to consider the first and second terms only. If α be the successive intervals of magnitude, the tabular second difference is

$$u(m+\alpha) + u(m-\alpha) - 2u(m) = \alpha^2 u''(m) \text{ approximately.}$$

For

$$u(m+\alpha) = u(m) + \alpha u'(m) + \frac{\alpha^2}{2}u''(m) \text{ approximately,}$$

and

$$u(m-\alpha) = u(m) - \alpha u'(m) + \frac{\alpha^2}{2}u''(m) \qquad \text{\textit{,,}}$$

$$\frac{1}{2h} = 1.046 \times \text{probable error.}$$

Thus the approximate correction is

$$-\left(\frac{1.046 \times \text{probable error}}{\text{tabular interval}}\right)^2 \times \text{tabular second difference.}$$

29. The Precision of the Probable Error deduced from $r = 0.6745\sqrt{\dfrac{[vv]}{n-1}}$. Effect of Random Sampling.

In § 23 we derived certain expressions for the residuals of n observations in terms of the true errors of the observations. If each of these expressions be squared, and the results added together, we obtain the expression

$$[vv] = \frac{1}{n^2}\{n(n-1)[\epsilon\epsilon] - 2n[\epsilon_s\epsilon_t]\},$$

where $[\epsilon_s\epsilon_t]$ denotes the sum of the products of all possible pairs from among the n quantities $\epsilon_1, \epsilon_2, \ldots, \epsilon_n$.

The equation may be written

$$\frac{[vv]}{n-1} = \frac{[\epsilon\epsilon]}{n} - \frac{2}{n(n-1)}[\epsilon_s\epsilon_t].$$

If we formed the equation for a large number of samples

of n observations from among a very large number of observations, and took the average value of each side of the equation, we should obtain

$$\frac{[vv]}{n-1} = \mu^2 - \text{mean value of } \epsilon_s \epsilon_t.$$

The derivation of the usual formula for μ^2 seems therefore to be equivalent to regarding the mean value of $\epsilon_s \epsilon_t$ as zero. That this is justifiable can easily be seen from § 20, equation (4'), where it was found that the mean value of the product $V_s V_t$, for all possible products among N residuals, is equal to $-\frac{\mu^2}{N-1}$. When the number of observations N is sufficiently large, it was seen that the V's could be regarded as the true errors ϵ, so that we may say that

$$\text{the mean value of } \epsilon_s \epsilon_t = -\frac{\mu^2}{N-1}.$$

When the number of observations N is very large, this quantity becomes vanishingly small, and may be neglected. Our equation above may then be written

$$\frac{[vv]}{n-1} = \mu^2.$$

This equation yields the mean result for a large number of cases. But in an individual set of n observations it will not be strictly true, and $\sqrt{\frac{[vv]}{n-1}}$ will not be the accurate M.S.E. of an infinite series of observations of which the given n observations form a random sample.

The error made in estimating μ^2 from the residuals in any particular case is

$$\frac{[vv]}{n-1} - \mu^2 = \frac{[\epsilon\epsilon]}{n} - \frac{2}{n(n-1)}[\epsilon_s \epsilon_t] - \mu^2.$$

The square of this error is

$$\frac{[\epsilon\epsilon]^2}{n^2} + \frac{4}{n^2(n-1)^2}[\epsilon_s \epsilon_t]^2 - \frac{4}{n^2(n-1)}[\epsilon\epsilon][\epsilon_s \epsilon_t] + \mu^4$$

$$- 2\frac{[\epsilon\epsilon]}{n}\mu^2 + \frac{4}{n(n-1)}[\epsilon_s \epsilon_t]\mu^2.$$

If we take the mean value of this expression for a large

number of cases, the result will be the (M.S.E.)² of μ^2. The mean values of all the terms in the expression above will be considered in turn:

$$[\epsilon\epsilon]^2 = [\epsilon^4] + 2[\epsilon_s^2\epsilon_t^2],$$

and

$$[\epsilon^4] = n \times \text{mean value of } \epsilon^4 = \frac{2nh}{\sqrt{\pi}} \int_0^\infty x^4 e^{-h^2x^2} dx = \frac{3n}{4h^4} = 3n\mu^4,$$

$$2[\epsilon_s^2\epsilon_t^2] = 2 \times \frac{n(n-1)}{2} \times \text{mean value of } \epsilon_s^2\epsilon_t^2$$

$$= n(n-1)\mu^4;$$

therefore   mean value of $\dfrac{[\epsilon\epsilon]^2}{n^2} = \dfrac{3\mu^4}{n} + \dfrac{n-1}{n}\mu^4.$

Again

$$[\epsilon_s\epsilon_t]^2 = [\epsilon_s^2\epsilon_t^2] + \text{terms involving first powers of } \epsilon.$$

When the mean value is formed the terms involving first powers vanish, therefore

$$\text{mean value of } \frac{4[\epsilon_s\epsilon_t]^2}{n^2(n-1)^2}$$

$$= \frac{4}{n^2(n-1)^2} \times \frac{n(n-1)}{2} \times \text{mean value of } \epsilon_s^2\epsilon_t^2$$

$$= \frac{2}{n(n-1)}\mu^4.$$

The mean value of the third term vanishes on account of the factor $[\epsilon_s\epsilon_t]$. The mean value of the fifth term is $-2\mu^4$, and the sixth term vanishes on account of the factor $[\epsilon_s\epsilon_t]$.

Thus we obtain the equations

$$(\text{M.S.E.})^2 \text{ of } \mu^2 = \frac{3\mu^4}{n} + \frac{n-1}{n}\mu^4 + \frac{2}{n(n-1)}\mu^4 - \mu^4$$

$$= \frac{2\mu^4}{n-1},$$

$$\text{M.S.E. of } \mu^2 = \mu^2 \sqrt{\frac{2}{n-1}}.$$

Since   M.S.E. of $\mu^2 = 2\mu \times$ M.S.E. of μ,

we finally obtain the result

$$\frac{\text{M.S.E. of } \mu}{\mu} = \frac{1}{\sqrt{2(n-1)}} = \frac{\cdot 707}{\sqrt{n-1}}.$$

Also, since $r = 0.6745\mu$, we have the relations

$$\frac{\text{M.S.E. of } r}{r} = \frac{.707}{\sqrt{n-1}}$$

and

$$\frac{\text{P.E. of } r}{r} = \frac{.4769}{\sqrt{n-1}},$$

where r is derived from the formula $r = 0.6745\sqrt{\dfrac{[vv]}{n-1}}$.

The following table gives the factor $\dfrac{.4769}{\sqrt{n-1}}$ for some different values of n.

n	$\dfrac{\text{P.E. of } r}{r}$	20 % error	50 % error
5	.288	.64	.24
10	.159	.40	.034
15	.127	.29	.008
20	.109	.21	.0002
30	.089	.12	.00014
40	.076	.076	8×10^{-6}
50	.068	.047	6×10^{-7}
60	.062	.030	5×10^{-8}
100	.048	.0050	

By the aid of this table, and the table of $\Theta(t)$ given in Appendix I, we may gain some idea of the number of observations which must be taken in order to yield values of the P.E. which are deserving of confidence. In the third and fourth columns above are given the probability that the P.E. should be 20 % out, and 50 % out, respectively. It is seen that with 10 observations the odds are only 3 to 2 that the calculated P.E. is within 20 % of the correct value, and about 30 to 1 that it is within 50 % of the correct value. The smallest number of observations whose P.E. shall be regarded as trustworthy will be to a certain extent a matter of individual opinion, depending upon the odds which the individual is prepared to regard as practically equivalent to certainty. But one can at any rate regard 50 as a number sufficient to yield a fairly reliable P.E.

When the number of observations is small, say 10, it is scarcely legitimate to regard the value of $0.6745 \sqrt{\dfrac{[vv]}{n-1}}$ as the P.E. of a single observation. It only yields an approximation, which cannot be regarded as a reliable one, to the value of the P.E.

There is another aspect of the question. The method of least squares is only strictly applicable to problems where the distribution of frequencies can be represented by a curve of the general form of the curve shown in fig. 3. The problem of determining the precision of the measurements is equivalent to determining the parameter h which defines the exact form of the curve. But the value of h, or of the P.E. r, must obviously cease to have its original meaning when the curve of frequencies is dissymmetrical, or is ill-defined on account of the smallness of the number of observations. An examination of the frequency distributions shown in figs. 1, 4, 5, 6 and 7, will show that it can scarcely be possible to gain any accurate idea of the true form of the frequency curve for a large number of observations from the chance distribution of a few observations.

It is only when the number of observations is large, say 50 or more, that the P.E. calculated from the formula

$$r = 0.6745 \sqrt{\frac{[vv]}{n-1}}$$

can be regarded as a reliable measure. If we use this formula to calculate r for a small number of observations, say 10, we cannot expect another observer, working under similar conditions, to obtain the same value of r for similar observations.

When the number of observations is small, or when the curve of frequencies is dissymmetrical (as in fig. 7), the calculated P.E. can only be regarded as some kind of measure of the mutual agreement of the observations in the series, a small P.E. indicating that the disagreement between individual observations is small.

30. A Comparison of the two Formulae for r.

It has been shown above that when r is calculated from the squares of the residuals, the P.E. of r is given by

$$\frac{\text{M.S.E. of } r}{r} = \frac{.707}{\sqrt{n-1}}.$$

Helmert* has shown, by an investigation which is too long and complicated to be included here, that when r is calculated from Peters' formula

$$r = 0.8453 \frac{v]}{\sqrt{n(n-1)}},$$

the M.S.E. of r is given by

$$\frac{\text{M.S.E. of } r}{r} = \sqrt{\frac{\pi - 2}{2(n-1)}} = \frac{\cdot 755}{\sqrt{n-1}}.$$

Thus the M.S.E. or P.E. of r calculated from the squares of the residuals is less than that of r calculated from the sum of the residuals in the ratio ·707 : ·755 or 1 : 1·07. The first method of calculating r is thus slightly better than the second, but the difference in their precision is not sufficiently great to justify the entire use of the first formula and the neglect of Peters' formula. Peters' formula has a very great advantage in that the labour involved in its use is by far less, particularly when the number of observations is great. And it is generally found that when the number of observations is fairly large, and the curve of errors is symmetrical the two formulae for r yield almost identical results. The two formulae have an equally strong theoretical basis, and when the results disagree, it is because the frequency distribution does not follow the normal law; and in this case the method of least squares should be applied with considerable caution.

The following practical consideration is very important. In a long series of observations it often happens that one or two observations are rejected as discordant. The retention of a discordant observation generally makes a considerable difference in the value of $\sqrt{[vv]}$, but a very much smaller difference in the value of $v]$. Thus when there are observations which we are doubtful about retaining, it is probably better to use Peters' formula.

* *Astron. Nachrichten*, Bd. 88, No. 2096-7.

CHAPTER IV

OBSERVATIONS OF DIFFERENT WEIGHT

31. The Weighting of Observations.

We have hitherto regarded all our observations as having equal precision, or, as is more generally said, as having equal weight. It is now necessary to consider how our formulae must be modified when the observations are regarded as having unequal weight, and consequently as having unequal importance in the determination of the most plausible value of the unknown. The meaning of "weight" and its effect upon the least squares solution can be most clearly seen by the consideration of a simple example.

Let x_1, x_2, x_3, x_4 be four observed values of an unknown quantity x. There are four equations of condition:

$$x - x_1 = v_1,$$
$$x - x_2 = v_2,$$
$$x - x_3 = v_3,$$
$$x - x_4 = v_4.$$

The most plausible value of the unknown is

$$x = \frac{x_1 + x_2 + x_3 + x_4}{4}.$$

Now suppose the first three observations to be grouped together, yielding a value $x'\left(= \frac{x_1 + x_2 + x_3}{3}\right)$ of the unknown for that group. Then there will be only two equations of condition:

$$x - x' = v',$$
$$x - x_4 = v_4.$$

Since x' is the mean of three determinations, while x_4 is a single determination, we may say that x' has a *weight* 3, while x_4 has unit or standard weight. The adopted value of the unknown may now be written

$$x = \frac{x_1 + x_2 + x_3 + x_4}{4} = \frac{3x' + x_4}{3 + 1}.$$

If x' and x_4 were direct observations of such a nature that it could be decided from any consideration that three observations such as x_4 must be taken and combined in order to yield a result as valuable as x', we could still say that x' should have a weight 3, while x_4 had the standard or unit weight. The adopted value of the unknown would still be written

$$x = \frac{3x' + x_4}{3 + 1}$$

The most plausible value of the unknown is derived by multiplying each observed value by its weight, adding together the products, and dividing the results by the sum of the weights.

The result may be generalised for any number of observations, with any assigned weights. If x_1, x_2, ..., x_n be n observations, whose weights are p_1, p_2, ..., p_n, respectively, the adopted value of the unknown x is

$$x = \frac{p_1 x_1 + p_2 x_2 + \ldots + p_n x_n}{p_1 + p_2 + \ldots + p_n} = \frac{[px]}{[p]}.$$

The adopted value of the unknown is called the "weighted mean." It may be noted in passing that this result is not altered when all the weights are increased or decreased in the same ratio.

In the simple case considered above, all that is meant by the weight 3 assigned to x' is that, on the average, three observations of unit weight must be combined in order to yield a result as good as x'. Similarly, in the general case, the meaning of a weight p_r assigned to an observation x_r, is that p_r observations of unit weight must be combined in order to yield as reliable a result as x_r. Later on we shall have to consider the different methods of assigning weights to observations, but for the present we are only concerned with the modifications in the methods of solution produced by the difference in weights.

The observational equations for a series of n weighted observations may be written

$$
\begin{aligned}
x_1 - x &= v_1 \quad \text{weight } p_1 \\
x_2 - x &= v_2 \quad \text{,, } \quad p_2 \\
&\cdots\cdots\cdots\cdots\cdots\cdots \\
&\cdots\cdots\cdots\cdots\cdots\cdots \\
x_n - x &= v_n \quad \text{,, } \quad p_n
\end{aligned}
\right\} \quad\cdots\cdots\cdots (A).
$$

Let the M.S.E. of a hypothetical observational equation of unit weight be μ. Then it follows from the definition of weight, that the mean values of the residuals are given by

$$
v_1{}^2 = \frac{\mu^2}{p_1}, \quad v_2{}^2 = \frac{\mu^2}{p_2}, \quad \text{etc.,}
$$

or
$$
\mu^2 = p_1 v_1{}^2 = p_2 v_2{}^2 = \ldots = p_n v_n{}^2.
$$

From this it follows that the above equations of condition are reduced to equations of equal M.S.E., and therefore to equations of equal weight, when each equation is multiplied by the square root of the corresponding weight. The system of equations may then be written

$$
\begin{aligned}
\sqrt{p_1}\,(x_1 - x) &= \sqrt{p_1} \cdot v_1 \\
\sqrt{p_2}\,(x_2 - x) &= \sqrt{p_2} \cdot v_2 \\
\text{etc.}
\end{aligned}
\right\} \quad\cdots\cdots\cdots (B).
$$

As all these equations have equal weight, the system may be solved by the ordinary method of least squares. The value of the unknown is obtained by making $\Sigma\,(pv^2)$ a minimum; i.e. by making

$$
p_1 (x_1 - x)^2 + p_2 (x_2 - x)^2 + \ldots \text{etc.}
$$

a minimum. Differentiating with respect to x, we obtain for the value of the weighted mean

$$
x = \frac{p_1 x_1 + p_2 x_2 + \ldots}{p_1 + p_2 + \ldots} = \frac{[px]}{[p]},
$$

which agrees with the value previously derived.

The equations (B) above are all of unit weight, and therefore the true M.S.E. μ of any one of them is given by

$$
\mu = \sqrt{\frac{[pvv]}{n-1}}.
$$

It follows that if r_0 be the P.E. of a single equation of unit weight,

$$r_0 = 0.6745 \sqrt{\frac{[pvv]}{n-1}}, \quad \text{or} \quad r_0 = 0.8453 \frac{\sqrt{p} \cdot v]}{\sqrt{n(n-1)}}.$$

The P.E. of an observation of weight p is $\dfrac{r_0}{\sqrt{p}}$; since such an observation is equivalent to p observations of unit weight and P.E. r_0 *. The weighted mean is equivalent to the arithmetic mean of $[p]$ observations, and so its P.E. r is given by

$$r = \frac{r_0}{\sqrt{[p]}} = 0.6745 \sqrt{\frac{[pvv]}{[p](n-1)}}.$$

It was assumed in the original definition of weight that the p's were integral. It should be noted, however, that the values of both the weighted mean and its P.E. are independent of the actual values of the weights, and depend only on their relative values.

32. Methods of Weighting.

(a) Arbitrary Scales.

It sometimes happens that the external conditions vary irregularly during a series of observations, in such a way that, although the effect upon the separate observations cannot be evaluated, yet the observer is able to decide that some of the observations are less affected than others. In such a case it appears legitimate to attach greater importance to the less affected observations. This is done by assigning to these observations a relatively higher weight than the more affected observations, the relative values of the weights being determined by the observer according to some arbitrary scale which he sets up for himself. Thus an astronomical observer making a long series of observations extending over many nights may fairly attach different weights to the results of the separate nights according to the steadiness or unsteadiness of seeing. When once these weights have been assigned, the weighted mean and its P.E. can be immediately evaluated by the use of the formulae deduced above.

---

* This result may also be derived as follows. Since r_0 is the P.E. of $\sqrt{p_r}v_r$, the P.E. of v_r is $\dfrac{r_0}{\sqrt{p_r}}$.

The greatest disadvantage of this method of weighting lies in its arbitrary and personal nature, as, in general, two observers making precisely the same series of observations would not assign the same relative weights to the separate observations. Further, a computer reducing a set of observations may legitimately reject the weights assigned by the observer when he considers that the observations are more affected by other factors not considered by the observer, than by the factors on which the weights are based. Thus, if an observer assigns weights to observations of the moon according to the steadiness of seeing, the computer may reject these weights if he finds that the observations are affected by the inequalities of the moon's limb to a greater extent than by the differences in seeing.

It is impossible to give any rules for the guidance of the inexperienced observer as to the formation of a scale of weights to represent the varying conditions during a series of observations. His safest plan is to observe only when external conditions are fairly stable, and to assign the same weights to all his observations, unless he has some very good reason for doing otherwise.

The general tendency of most observers is to overdo this type of weighting; i.e. to regard the bad observations as worse than they really are. "It appears that the longer time one is compelled to bestow, and does bestow, upon observations made under less favourable circumstances, in a great measure compensates external disadvantages; and that causes of errors of observation of which the observer himself has not been conscious, often influence him no less than those which obtrude themselves upon him*."

(b) Number of Observations in Grouped Means.

When it is required to combine a number of quantities each of which is the mean of a group of observations, nothing being known of the individual observations in each group, each group-mean is given a weight proportional to the number of observations in the group. The weighted mean thus obtained is clearly the same as the mean of all the observations. But the P.E. derived

---

* Ordnance Survey. Principal Triangulation.

from grouped means may differ slightly from that derived from the *actual* observations.

(c) Weighting by Probable Errors*.

Let x_1, x_2, ..., x_n be separate determinations of a quantity x, where the P.E. of x_s $(s = 1, 2, ..., n)$ is known. The most probable value of x is obtained by making

$$\Sigma h_s^2 (x_s - x)^2 \text{ a maximum.}$$

This is equivalent to making

$$\Sigma \frac{(x_s - x)^2}{r_s^2} \text{ a maximum.}$$

Differentiating with respect to x, we obtain

$$x = \frac{\left[\dfrac{1}{r_s^2} \cdot x_s\right]}{\left[\dfrac{1}{r_s^2}\right]} = \frac{[p_s x_s]}{[p_s]},$$

where

$$p_s = \frac{1}{r_s^2}.$$

This result may be interpreted as follows:

If it is necessary to combine a number of separate determinations of an unknown quantity, where the P.E. of each separate determination is known, the best result is obtained by assigning to each determination a weight inversely proportional to the square of its P.E.

This method of weighting is undoubtedly the best when it is possible to obtain trustworthy P.E.'s of the separate determinations, the number of observations on which each determination is based being not too small. When the number of observations is small there is always a danger of a run of luck causing the observations in a group to fall close together, so yielding a very small P.E., and consequently a very large weight. In such a case the computer must decide whether the small P.E. represents a true P.E. or is small simply through the accidentally close agreement of the observations in a group.

---

* In all that follows the M.S.E. μ may be used instead of the P.E. r, as the basis of weighting.

The following description of a practical method of weighting, extracted from Wright and Hayford's *Adjustment of Observations* (page 76), is particularly instructive. "A long-continued series of observations will show the kind of work an instrument is capable of doing under favourable conditions; and if work is done only when the conditions are favourable, the P.E. derived from a certain number of results will generally fall within limits that can be assigned *a priori*. For example, with the Lake Survey primary theodolites, which read to single seconds, the tenths being estimated, the work of several seasons showed that the mean of from 16 to 20 results of the value of a horizontal angle, each result being the mean of a reading with telescope direct and a reading with telescope reverse, need not be expected to be greater than 0″·3. If, therefore, after having measured a series of angles in a triangulation net with these instruments, the P.E.'s all fell within ± 0″·3, it was considered sufficient to assign to each angle the same weight."

No general rules can be given as to the best methods of assigning weights. The computer must take into consideration all the information available concerning the external conditions at the time of observation, the possible presence of constant errors, the type of instrument used, the reputation of the observer for accurate work, as well as the computed P.E. With all these in mind he must use his own judgment as to the best method of procedure. The inexperienced computer must guard against assigning widely divergent weights to his observations, unless he has very strong grounds for so doing.

In assigning weights according to P.E. it is necessary to consider the possibility of a systematic or constant error being present in the set of observations, causing an error in the final result. For, since the P.E. only takes account of the accidental errors, it can only be regarded as a valid measure of the precision of a result when it is certain that the result is not affected by systematic errors. Thus in an attempt to combine the values of the solar parallax obtained by different methods, we must consider not only the calculated P.E., but also the possible sources of systematic errors. Newcomb states that the errors principally to be feared in a determination of the solar parallax are not the accidental errors treated by the method of least squares, but the systematic ones arising principally from personal equation and an imperfect reduction of the observations to the centre of the planet, or to the sun. It is therefore useless to assign relative weights based on the P.E.'s of the separate determinations. From a careful study of the systematic errors entering into the different estimates of the solar parallax, Newcomb assigned the weights shown in the table below. These weights are not strictly proportional to the calculated P.E.'s.

	Parallax in Seconds	Weights
Gill's observations of Mars	8·780 ± 0·020	1
Contact observations of transit of Venus	8·794 ± 0·018	1
Aberration and velocity of light	8·798 ± 0·005	16
Parallactic inequality of the moon ...	8·799 ± 0·007	5
Minor planets (Gill)	8·807 ± 0·007	8
Leverrier's method	8·818 ± 0·030	0·5
Venus...	8·857 ± 0·022	1

33. An alternative Method of Evaluating the Precision of the Weighted Mean when Weighting according to Probable Errors.

It has been shown above that if we wish to combine a number of determinations of any quantity when the P.E. of each separate determination is known, the weight to be given to each determination must be proportional to the inverse square of the probable error. If $x_1, x_2, \ldots, x_n$ be the n determinations, whose P.E.'s are $r_1, r_2, \ldots, r_n$, the weight p_s of the determination x_s is given by

$$p_s = \frac{1}{r_s^2}.$$

The weighted mean can be evaluated by means of the equation

$$x = \frac{\left[\dfrac{x_s}{r_s^2} \right]}{\left[\dfrac{1}{r_s^2} \right]} \quad \ldots\ldots\ldots\ldots\ldots\ldots\ldots(1).$$

The P.E. of the weighted mean can be evaluated from the residuals by the use of the formula

$$r = 0.6745 \sqrt{\frac{[pvv]}{[p](n-1)}} \quad \dots\dots\dots\dots(2).$$

But there is another way of approaching the question. The weighted mean x, defined by the equation

$$x = \frac{\left[\dfrac{x_s}{r_s^2}\right]}{\left[\dfrac{1}{r_s^2}\right]},$$

is a linear function of n quantities x_1, x_2, etc.. whose P.E.'s are known. Hence if r be the P.E. of x,

$$r^2 = \frac{\dfrac{1}{r_1^4}r_1^2 + \dfrac{1}{r_2^4}r_2^2 + \dots}{\left(\dfrac{1}{r_1^2} + \dfrac{1}{r_2^2} + \dots\right)^2},$$

or

$$\frac{1}{r^2} = \frac{1}{r_1^2} + \frac{1}{r_2^2} + \dots + \frac{1}{r_n^2} \quad \dots\dots\dots\dots(3).$$

The results derived from equations (2) and (3) will in general differ. For the first of these bases the calculation of r upon the differences between the individual determinations x_1, x_2, etc.; while the second method neglects entirely these differences. There naturally arises the question, "which of the two methods of evaluating r is the better?" The answer to this question depends to some extent upon the material under consideration. If the differences between the individual determinations which have to be combined are attributable to systematic errors entering into different determinations in different ways, it is clear that the P.E. of a determination can give no clear estimate of the reliability of that determination. In such a case equation (3) yields a value of r which is of no use as a measure of the reliability of the result. We are then forced to use equation (2). If the number of determinations to be combined be small, say, 3 or 4, and we have no reason to suspect the presence of systematic errors, it is better to

use equation (3).   In all other cases it is safer to use equation (2);
i.e. to calculate the P.E. from the residuals.

EXAMPLES.

1.   Given the following six determinations of the parallax of the star
Lalande 21185, find the weighted mean and its P.E.

Parallax	Weight (p)	x	px	v	v^2	pv^2
$0''{\cdot}507$	8	107	856	104	10816	86528
·438	5	38	190	35	1225	6125
·381	2	−19	−38	−22	484	968
·371	8	−29	−232	−32	1024	8192
·350	13	−50	−650	−53	2809	36517
·402	20	2	40	−1	1	20
	56		166			138350

Let the parallax of the star be $0''{\cdot}4 + x'' \times 10^{-3}$ Then the six values of x
are given in the third column.   Each value of x is multiplied by the corresponding weight, and the result written in the fourth column.   The weighted
mean gives for x,

$$x = \tfrac{166}{56} = 3.$$

The adopted parallax is therefore $0''{\cdot}403$.
The residual v obtained by subtracting the weighted mean from x is
written in the fifth column, v^2 in the sixth column, and pv^2 in the seventh
column.   The sum of the last column yields $[pvv]$.
The P.E. of the weighted mean is

$$0{\cdot}6745 \sqrt{\frac{[pvv]}{[p](n-1)}} = 0{\cdot}6745 \sqrt{\frac{138350}{56 \times 5}} = 15{\cdot}0$$

in units of the third decimal place.
The final result may be written

$$0''{\cdot}403 \pm {\cdot}015.$$

2. Find the weighted mean and its P.E. for the following determinations of the difference of longitude between two places :

	Weights (p)	x	px	v	v^2	pv^2
19^m 1^s·42 ± 0^s·044	1·1	− 3	− 3·3	− 4	16	17·6
·37 ± ·037	1·6	− 8	− 12·8	− 9	81	129·6
·38 ± ·036	1·7	− 7	− 11·9	− 8	64	108·8
·45 ± ·036	1·7	0	0	− 1	1	1·7
·60 ± ·046	1·0	15	15·0	14	196	196·0
·55 ± ·045	1·0	10	10·0	9	81	81·0
·57 ± ·047	1·0	12	12·0	11	121	121·0
	9·1		9·0			655·7

The weights of the determinations are proportional to

$$\frac{1}{(44)^2}, \quad \frac{1}{(37)^2}, \quad \frac{1}{(36)^2}, \quad \frac{1}{(36)^2}, \quad \frac{1}{(46)^2}, \quad \frac{1}{(45)^2}, \quad \frac{1}{(47)^2}.$$

Let the weight of the last determination be made unity. Then the others become

$$\left(\frac{47}{44}\right)^2, \quad \left(\frac{47}{37}\right)^2, \quad \left(\frac{47}{36}\right)^2, \quad \left(\frac{47}{36}\right)^2, \quad \left(\frac{47}{46}\right)^2, \quad \left(\frac{47}{45}\right)^2, \quad 1.$$

or   1·1,   1·6,   1·7,   1·7,   1·0,   1·0,   1·0.

These weights are written down in the second column of the table. The longitude difference is assumed to be 19^m 1^s·45 + x × 0″·01. The separate determinations of x are written in the third column, and the values of px in the fourth column. The sum of the fourth column, divided by the sum of the weights, gives $\frac{9\cdot0}{9\cdot1}$ or 1·0 for the weighted mean of x. The residual v is $x − 1$. The rest of the table is self-explanatory.

The P.E. of the weighted mean

$$= 0·6745 \sqrt{\frac{655·7}{9·1 \times 6}} = 2·3$$

in units of the second decimal place.

The final result may therefore be written

19^m 1^s·460 ± 0^s·023.

3. From the following determinations of the parallax of 61 Cygni find the weighted mean, and its P.E. :

$$0''\!\cdot\!316 \pm 0''\!\cdot\!016$$
$$\cdot 216 \pm \quad \cdot 029$$
$$\cdot 333 \pm \quad \cdot 035$$
$$\cdot 290 \pm \quad \cdot 009$$
$$\cdot 300 \pm \quad \cdot 007$$
$$\cdot 387 \pm \quad \cdot 015$$
$$\cdot 328 \pm \quad \cdot 029$$
$$\cdot 298 \pm \quad \cdot 005$$
$$\cdot 238 \pm \quad \cdot 020$$
$$\cdot 388 \pm \quad \cdot 017$$

[Give each determination a weight proportional to $\dfrac{1}{(\text{P.E.})^2}$, and evaluate P.E. of the weighted mean from the residuals.]

4. An angle was determined in three separate years, the following results being obtained :

$$149°\ 16'\ 51''\!\cdot\!48 \pm 0''\!\cdot\!45$$
$$48\ \cdot47 \pm 0\ \cdot28$$
$$49\ \cdot72 \pm 0\ \cdot25$$

Find the most probable measure of the angle.

[In this case the differences between the individual observations are greatly in excess of the P.E.'s. There are obviously systematic errors present, so that the P.E.'s do not represent the whole error. It is probably preferable in this case to adopt the simple A.M. as the value of the angle.]

5. If the probable accidental error of observing a star is $0''\!\cdot\!30$, and the probable quasi-systematic error of a gradation is $0''\!\cdot\!20$, how would you combine 7 observations of one star with 3 observations of another star?

Referring to page 51, we find that

$$(\text{P.E.})^2 \text{ of mean of 7 observations} = (\cdot 20)^2 + \tfrac{1}{7}\,(\cdot 30)^2,$$
$$(\text{P.E.})^2 \text{ of mean of 3 observations} = (\cdot 20)^2 + \tfrac{1}{3}\,(\cdot 30)^2.$$

6. From the table of determinations of the solar parallax given on page 67, using the weights assigned by Newcomb, show that the weighted mean is

$$8''\!\cdot\!802 \pm 0''\!\cdot\!005.$$

7. The difference between the observed values and the assumed value of an angle is given by the following set of determinations :

$$1''\!\cdot\!27 \pm 0''\!\cdot\!11 \text{ determined from 4 observations}$$
$$1\ \cdot00 \pm 0\ \cdot10 \qquad ,, \qquad ,, \quad 3 \quad ,,$$
$$0\ \cdot91 \pm 0\ \cdot07 \qquad ,, \qquad ,, \quad 4 \quad ,,$$
$$1\ \cdot12 \pm 0\ \cdot13 \qquad ,, \qquad ,, \quad 6 \quad ,,$$
$$1\ \cdot30 \pm 0\ \cdot11 \qquad ,, \qquad ,, \quad 6 \quad ,,$$
$$1\ \cdot42 \pm 0\ \cdot19 \qquad ,, \qquad ,, \quad 6 \quad ,,$$
$$1\ \cdot45 \pm 0\ \cdot15 \qquad ,, \qquad ,, \quad 6 \quad ,,$$

How would you combine these values to obtain the most probable value
(a) when the above probable errors are given simply by the discordances
between the individual observations of each group, and (b) when in each case
the probable error is determined from a long series of which each member of
the above groups is a sample ?

In the former case, determine the most probable value of the angle and
its probable error.

[In the first case the observations are to be weighted according to the
number of observations in each group, while in the second case the P.E.'s
must also be taken into account.]

MISCELLANEOUS EXAMPLES INVOLVING ONE UNKNOWN.

1.   If the error of a clock determined at time

$$0^h \text{ is } a_1{}^s \pm r_1,$$
$$12^h \text{ is } a_2{}^s \pm r_2,$$

find the clock error and its P.E. for an interpolated time 8^h.

2.   The equatorial velocity of the sun, determined by a spectroscopic
method, yielded the following results:

Element	Number of lines	Km./sec.
Fe	18	1·857
Ti	8	1·883
Cr	6	1·883
Sc	5	1·847
Ca	3	1·845
V	3	1·870
Zr	3	1·915
Mn	2	1·819
Mg	1	1·989
Ni	1	1·848

Combine these values (a) giving all elements equal weight, (b) giving each
element a weight equal to the number of lines measured.

3. $2n+1$ observations are made, each with the same P.E.   Show that the
probability of the error of the median being between x and $x+dx$ is

$$\frac{2n+1\,!}{n!\,n!}\left(\tfrac{1}{4}-a^2\right)^n da,$$

where
$$a=\frac{h}{\sqrt{\pi}}\int_0^x e^{-h^2 t^2}\,dt.$$

In the case of five observations, show that the chance of the error of the
median being numerically greater than r is $\frac{53}{256}$.

4. The probability of an event happening is p, so that in m trials the event happens on an average mp times. Find the M.S.E. of the number mp.

Let q be the probability of the event not happening, so that

$$q = 1 - p.$$

The whole series is given by

$$(p+q)^m = p^m + mp^{m-1} q + \ldots + q^m.$$

This may be interpreted as follows: The frequency of the event happening m times is p^m, of its happening $(m-1)$ times is $mp^{m-1}q$, etc. The separate terms of the Binomial Series give the frequency distribution of the different possible numbers of successes in m trials. Or if x be the number of times the event occurs in m trials, we have the following frequency distribution:

$$x = m, \qquad f = p^m,$$
$$x = m - 1, \qquad f = mp^{m-1}q,$$
$$x = m - 2, \qquad f = \frac{m \cdot m - 1}{1 \cdot 2} p^{m-2} q^2,$$
$$\text{etc.,}$$
$$x = 0, \qquad f = q^m.$$

Taking the origin at $x = m$, we find,

$$\text{mean value} = mp^{m-1} q + \frac{m \cdot m - 1}{1 \cdot 2} p^{m-2} q^2 + \ldots$$
$$= mq (p+q)^{m-1}$$
$$= mq.$$

$$\Sigma f (x-m)^2 = mp^{m-1} q + 2m (m-1) p^{m-2} q^2 + \ldots + m^2 q^m$$
$$= mp^{m-1} q + m (m-1) p^{m-2} q^2 + \ldots + mq^m$$
$$\qquad + m (m-1) p^{m-2} q^2 + \ldots + m (m-1) q^m$$
$$= mq + m (m-1) q^2.$$

But $(\text{M.S.E.})^2$
$$= \Sigma f (x - mp)^2 = \Sigma f (x-m)^2 - (\text{mean value})^2$$
$$= mq + m (m-1) q^2 - m^2 q^2 = mpq.$$

The use of the formula will be perhaps best shown by a simple application. It has been stated that in the British Isles the proportion of male to female children is 1050 to 1000. Hence the probability that a child should be male is $\frac{1050}{2050}$. Of 100,000 births the number of male children should be

$$\tfrac{105}{205} \times 100,000 \text{ or } 51,219.$$

The mean square error of this estimate is

$$\sqrt{100,000 \times \tfrac{105}{205} \times \tfrac{100}{205}} \text{ or } 158 \text{ approximately.}$$

If it were found that among 100,000 children born in this country, 51,500 were male, the deviation from the expected value would be only about twice the M.S.E., and need not be regarded as abnormal. But if among 10,000,000 children 5,150,000 were male, the deviation from the expected value would be

28,100, while the M.S.E. of the deviation would be 1,580. The actual deviation would be about 18 times the M.S.E., and we should conclude that the normal ratio of male to female children had been definitely changed.

5.   Given that the P.E. of a single observation is ·14, how many observations must be taken and combined in order that the P.E. of the mean shall be less than ·02?

6.   A series of 100 observations of an angle gives for the P.E. of a single observation 1″·75.   What is the probability that the error of the mean is not greater than 0″·25?

7.   An angle is measured 100 times, and the P.E. of a single observation is 1″·75.   How many errors will be greater than 0″·25 and less than 1″·25?

8.   If the P.E. of a single observation is 1·5, how many observations must be combined in order that the odds may be 3 to 1 that the mean is within ·25 of truth?

CHAPTER V

THE GENERAL PROBLEM OF THE ADJUSTMENT OF IN-
DIRECT OBSERVATIONS INVOLVING MORE THAN ONE
UNKNOWN QUANTITY

34. IN the cases hitherto considered, the problem has been to find the most probable value of an unknown quantity, given a number of direct observations of that quantity. The arithmetic mean was adopted as the best value of the unknown. We must now consider the case where the quantity measured is not itself the unknown whose value is required, but is expressible as a function (not of necessity linear) of a number of unknown quantities. The problem may be briefly stated thus: Given a number of measurements of certain functions of a number of unknowns, to find the values of the unknowns, and their probable errors.

Let the n observed quantities be $M_1, M_2, ..., M_n$; and let their unknown errors be $v_1, v_2, ..., v_n$. Then it is given that $M_1 + v_1$, $M_2 + v_2$, etc. can be accurately expressed as functions of the unknowns X, Y, Z, etc.,

$$\left.\begin{aligned} f_1(X, Y, Z, ...) &= M_1 + v_1 \\ f_2(X, Y, Z, ...) &= M_2 + v_2 \\ &\cdots\cdots\cdots\cdots\cdots\cdots\cdots\cdots \\ &\cdots\cdots\cdots\cdots\cdots\cdots\cdots\cdots \end{aligned}\right\} \quad\cdots\cdots\cdots\cdots(1).$$

There will be one equation of this form for each observation; and in the problems with which we shall have to deal, the number n of equations in (1) will be greater than the number m of unknowns.

Now suppose approximate values of the unknowns to be known, or to have been deduced by solving a sufficient number of the equations in (1). Let these approximate values be X_0, Y_0, Z_0, etc., and let

$$X = X_0 + x, \quad Y = Y_0 + y, \quad Z = Z_0 + z, \text{ etc.,}$$

where it may be assumed that the corrections x, y, z, etc are small, so that their squares may be neglected.

The first equation in (1) may then be written

$$f_1 (X_0, Y_0, Z_0, \ldots) + x \frac{\partial f_1}{\partial X_0} + y \frac{\partial f_1}{\partial Y_0} + z \frac{\partial f_1}{\partial Z_0} + \ldots = M_1 + v_1.$$

Let

$$\frac{\partial f_1}{\partial X_0} = a_1, \quad \frac{\partial f_1}{\partial Y_0} = b_1, \text{ etc.,}$$

$$\frac{\partial f_2}{\partial X_0} = a_2, \quad \frac{\partial f_2}{\partial Y_0} = b_2, \text{ etc.,}$$

$$\ldots\ldots\ldots\ldots\ldots\ldots\ldots\ldots\ldots,$$

$$\ldots\ldots\ldots\ldots\ldots\ldots\ldots\ldots\ldots;$$

$$+ M_1 - f_1 (X_0, Y_0, Z_0, \ldots) = l_1,$$

$$+ M_2 - f_2 (X_0, Y_0, Z_0, \ldots) = l_2,$$

$$\ldots\ldots\ldots\ldots\ldots\ldots\ldots\ldots\ldots\ldots,$$

$$\ldots\ldots\ldots\ldots\ldots\ldots\ldots\ldots\ldots\ldots$$

Then equations (1) may be written

$$\left. \begin{aligned} a_1 x + b_1 y + c_1 z + \ldots - l_1 &= v_1 \\ a_2 x + b_2 y + c_2 z + \ldots - l_2 &= v_2 \\ \ldots\ldots\ldots\ldots\ldots\ldots\ldots\ldots\ldots \\ \ldots\ldots\ldots\ldots\ldots\ldots\ldots\ldots\ldots \end{aligned} \right\} \quad \ldots\ldots\ldots\ldots(2),$$

where the a's, b's, c's, l's, etc. are known.

Equations (1) or (2) are called the "observational equations." The problem has now been reduced to the case where the equations are all linear. If the values of x, y, z, etc. obtained by solving equations (2) be small, we may rest content with our solution; but if some of them should be large, it may be necessary to repeat the solution, taking the results of the first solution as approximate values of X, Y, Z, etc. It is found in practice that the approximation step saves considerable labour even in cases where the original observational equations are strictly linear.

35. Formation of the Normal Equations.

If the observational equations all have the same weight, or are liable to the same mean error, the same discussion as for one unknown quantity will apply (see §§ 13 and 14). The probability of the occurrence of a residual v may be written

$$Ce^{-h^2 v^2}.$$

The probability of the coexistence of the system of residuals $v_1, v_2, \ldots, v_n$ may be written

$$C^n e^{-h^2 [vv]}.$$

The most probable values of the unknown will be such as to make this probability a maximum. The expression is greatest when $[vv]$ is least, and so the most probable values of the unknowns are given by the condition that

$$[vv] = \text{a minimum,}$$

or

$$\sum_{r=1}^{r=n} (a_r x + b_r y + \ldots - l_r)^2 = \text{a minimum.}$$

The conditions for a minimum are obtained by equating to zero the differential coefficients of this expression with respect to x, y, etc.,

$$\left.\begin{aligned}
a_1(a_1 x + b_1 y + \ldots - l_1) + a_2(a_2 x + b_2 y + \ldots - l_2) + \ldots = 0 \\
b_1(a_1 x + b_1 y + \ldots - l_1) + b_2(a_2 x + b_2 y + \ldots - l_2) + \ldots = 0 \\
\ldots\ldots\ldots\ldots\ldots\ldots\ldots\ldots\ldots\ldots\ldots\ldots\ldots\ldots\ldots \\
\ldots\ldots\ldots\ldots\ldots\ldots\ldots\ldots\ldots\ldots\ldots\ldots\ldots\ldots\ldots
\end{aligned}\right\} (3).$$

Collecting coefficients, we may write these equations in the form

$$\left.\begin{aligned}
[aa]x + [ab]y + [ac]z + \ldots - [al] = 0 \equiv \xi \\
[ab]x + [bb]y + [bc]z + \ldots - [bl] = 0 \equiv \eta \\
[ac]x + [bc]y + [cc]z + \ldots - [cl] = 0 \equiv \zeta \\
\ldots\ldots\ldots\ldots\ldots\ldots\ldots\ldots\ldots\ldots\ldots\ldots \\
\ldots\ldots\ldots\ldots\ldots\ldots\ldots\ldots\ldots\ldots\ldots\ldots
\end{aligned}\right\} \ldots\ldots(4).$$

These equations are called the *normal equations*. There is one normal equation corresponding to each unknown, and our problem has therefore been reduced to that of solving a set of linear equations whose number is the same as the number of unknowns. The solution of the normal equations gives the most probable values of the corrections x, y, z, etc., and from these corrections the values of the original unknowns X, Y, Z, etc. can be immediately deduced.

Since the coefficients of the normal equations are symmetrical about the principal diagonal, it is convenient to write the normal equations in the following abbreviated form, with half the cross products omitted,

$$\left.\begin{aligned}
[aa]\,x + [ab]\,y + [ac]\,z + \ldots - [al] &= 0 \\
+ [bb]\,y + [bc]\,z + \ldots - [bl] &= 0 \\
+ [cc]\,z + \ldots - [cl] &= 0 \\
\ldots\ldots\ldots\ldots\ldots\ldots\ldots\ldots\ldots\ldots \\
\ldots\ldots\ldots\ldots\ldots\ldots\ldots\ldots\ldots\ldots
\end{aligned}\right\} \quad\ldots\ldots\ldots(5).$$

There are m normal equations when the number of unknowns X, Y, Z, etc. is m. The total number of products to be estimated is $\frac{1}{2}m\,(m+3)$, made up of m products $[aa]$, $[bb]$, etc., m products $[al]$, $[bl]$, etc., and $\frac{1}{2}m\,(m-1)$ products $[ab]$, $[bc]$, $[ca]$, etc.

Equations (3) may be written

$$[av] = 0 = [bv] = [cv] = \ldots \quad\ldots\ldots\ldots\ldots(6),$$

corresponding to the relation $[v] = 0$ obtained in the case of one unknown quantity. These relations afford a check upon the values of v_1, v_2, etc. obtained from the solution of the equations.

If the observational equations are liable to different mean errors, or in other words have different weights p_1, p_2, etc., the equations (2) above must be multiplied by $\sqrt{p_1}$, $\sqrt{p_2}$, etc., respectively, so as to reduce them to equations of equal weight. *Save for this multiplication it is not permissible to multiply any observational equation by an arbitrary factor.* It is then necessary to make

$$[pvv]\text{ a minimum.}$$

This condition leads to the set of normal equations

$$\left.\begin{aligned}
[paa]\,x + [pab]\,y + [pac]\,z + \ldots - [pal] &= 0 \\
[pab]\,x + [pbb]\,y + [pbc]\,z + \ldots - [pbl] &= 0 \\
\ldots\ldots\ldots\ldots\ldots\ldots\ldots\ldots\ldots\ldots \\
\ldots\ldots\ldots\ldots\ldots\ldots\ldots\ldots\ldots\ldots
\end{aligned}\right\} \quad\ldots(4'),$$

with the conditions

$$[pav] = [pbv] = \ldots = 0 \quad\ldots\ldots\ldots\ldots(6').$$

If we slightly revise our notation so that

$$[aa] \text{ stands for } \Sigma pa^2,$$
$$[ab] \quad ,, \quad ,, \quad \Sigma pab,$$
$$\text{etc.,}$$

then equations (4') and (6') are included in equations (4) and (6), and the same method of solution can be applied in each case. In all that follows we shall take account of possible differences in weight by regarding $[aa]$, $[ab]$, etc. as Σpa^2, Σpab, etc.

36. Independence of the Normal Equations.

If the normal equations are not all independent, it will be possible to derive any one of them by combining some or all of the others, and the solution of equations (4) will not lead to a determinate solution for x, y, z, etc. In this case it must be possible to find m constants, the ratios of $f, g, h, \ldots, k$, so that

$$f\xi + g\eta + h\zeta + \ldots + k = 0,$$

where ξ, η, ζ, etc. are written for the normal equations (4), and m is the number of unknowns x, y, z, etc.

Collecting the coefficients of x, y, z, in the last equation, we find

$$\left.\begin{aligned}
f[aa] + g[ab] + \ldots &= 0 \\
f[ab] + g[bb] + \ldots &= 0 \\
\ldots\ldots\ldots\ldots\ldots\ldots\ldots & \\
\ldots\ldots\ldots\ldots\ldots\ldots\ldots &
\end{aligned}\right\} \ldots\ldots\ldots\ldots(7).$$

and finally, $\quad f[al] + g[bl] + \ldots = -k$

Multiplying the first m of these equations by f, g, h, etc. respectively, and adding the products, we find

$$\sum_{r=1}^{r=n} (fa_r + gb_r + \ldots)^2 = 0,$$

or

$$\left.\begin{aligned}
fa_1 + gb_1 + \ldots &= 0 \\
fa_2 + gb_2 + \ldots &= 0 \\
\ldots\ldots\ldots\ldots\ldots & \\
\ldots\ldots\ldots\ldots\ldots &
\end{aligned}\right\} \ldots\ldots\ldots\ldots(8).$$

There will be one of these equations for each observational

equation. Multiplying equations (8) by l_1, l_2, etc. and adding, we find

$$f[al] + g[bl] + \ldots = 0 = -k \text{ from (7).}$$

It follows from equations (8) that there cannot be more than $(m-1)$ independent observational equations; for equations (8) will yield a relation between the coefficients of any m of these equations. The problem of solving for x, y, z, etc. is then indeterminate from the beginning.

The above discussion shows that if the normal equations are not independent, there could not have been m independent observational equations. We infer that if there are at least m independent observational equations, there will result m independent normal equations, the solution of which yield determinate values of the m unknowns. This may be otherwise stated as follows,— If there are sufficient observational equations to determine or over-determine the solution, the normal equations yield a determinate solution.

The condition of independence is in general satisfied in the problems which arise in practice. We can then proceed to the formation and solution of the normal equations.

37. Checks on the Formation of the Normal Equations.

Before the normal equations can be written in a complete form, it is necessary to compute the products $[aa]$, $[bb]$, $[ab]$, $[al]$, etc. It is necessary to check the formation of these products. The most convenient form of check is the following.

Let
$$a_r + b_r + \ldots + l_r = s_r.$$

Then multiplying each s-equation by the corresponding a, and adding, we find

$$[aa] + [ab] + \ldots + [al] = [as].$$

And similarly

$$[ab] + [bb] + \ldots + [bl] = [bs],$$
$$\text{etc.}$$

Each of these equations is a check on the sum of the coefficients of one of the normal equations. The calculation of $[as]$, $[bs]$, etc. yields a double check on each of the cross-products. The additional work involved is very slight. We evaluate s_r for each of the observational equations, and in the subsequent work treat s_r in the same way as any other coefficient.

The work of computing the products is carried out with the aid of tables of squares, of products, or of logarithms, or with an arithmometer, a machine for performing multiplication and division. If the work is to be carried on by the aid of tables of squares, the products $[ab]$ etc. can be derived as follows:

$$(a+b)^2 = a^2 + b^2 + 2ab;$$
$$\therefore \; 2[ab] = [(a+b)^2] - [aa] - [bb].$$

Since $[aa]$ and $[bb]$ have to be evaluated in any case, the evaluation of $[ab]$ is simply replaced by the evaluation of $[(a+b)^2]$.

In any extensive piece of work involving a large number of observational equations, or a large number of unknowns, it is a great saving of labour to carry on the work in a fixed form, so that the whole of the work, including the application of the checks, shall be as uniform and mechanical as possible. It will also be necessary to have a system of checks at different stages in the process of solving the normal equations, so that any arithmetical error shall be immediately detected.

A great deal of labour can be saved in some cases by selecting new units of the unknowns, in such a way that the coefficients in the equations shall all be of the same order of magnitude.

38. It is sometimes possible to effect a considerable simplification of the work of solution by means of a simple substitution; e.g. when the observational equations are of the form

$$ax + bxy = l,$$

the solution is facilitated by the substitution

$$u = xy.$$

The observational equations then become

$$ax + bu = l,$$

and the solution can be carried out in the usual way. This method of introducing a new unknown dispenses with the approximation which would otherwise be necessary in the first step of the solution, by reducing the equations of conditions to a linear form.

It is *not* however permissible to calculate the probable errors of x and u in the following manner:

$$y = \frac{u}{x},$$

$$\therefore \; dy = \frac{du}{x} - \frac{u}{x^2}\, dx,$$

$$r_y^2 = \frac{r_u^2}{x^2} + \frac{y^2}{x^2}\, r_x^2.$$

For such a procedure assumes that x and u are independent variables, and that the errors of x and u are independent. (See p. 36, § 22, and p. 116, § 5?.)

39. The Formation of the Normal Equations.

Example 1.   Given the four equations*,

$$x - y + 2z - 3 = 0,$$
$$3x + 2y - 5z - 5 = 0,$$
$$4x + y + 4z - 21 = 0,$$
$$-x + 3y + 3z - 14 = 0,$$

it is required to find the most probable values of x, y, z.

If we solve the first three of these equations, we find

$$x = 2\tfrac{4}{7}, \quad y = 3\tfrac{2}{7}, \quad z = 1\tfrac{6}{7}.$$

Substituting these values in the last equation, we find

$$-x + 3y + 3z - 14 = -\tfrac{8}{7},$$

and so the fourth equation is not satisfied. We therefore proceed to form the normal equations.

The formation of the normal equations is best carried out in tabular form. The coefficients of the observational equations are first written down, and the products formed in turn.

a	b	c	l	s
1	−1	2	3	5
3	2	−5	5	5
4	1	4	21	30
−1	3	3	14	19

aa	ab	ac	al	as
1	−1	2	3	5
9	6	−15	15	15
16	4	16	84	120
1	−3	−3	−14	−19
·27	6	0	88	121

* Gauss, *Theoria Motus*, § 184.

ab	bb	bc	bl	bs
	1	−2	−3	−5
	4	−10	10	10
	1	4	21	30
	9	9	42	57
6	15	1	70	92

ac	bc	cc	cl	cs
		4	6	10
		25	−25	−25
		16	84	120
		9	42	57
0	1	54	107	162

al	bl	cl	ll	ls
			9	15
			25	25
			441	630
			196	266
88	70	107	671	936

Checks are applied to all the sums as they are formed.

The normal equations may then be written

$$27x + 6y \qquad = 88,$$
$$6x + 15y + z = 70,$$
$$y + 54z = 107.$$

No special method is necessary for the solution of these equations. Substituting for x and z in the second equation, we find

$$\frac{6}{27}(88 - 6y) + 15y + \frac{1}{54}(107 - y) = 70,$$

giving $\qquad\qquad y = 3\cdot55.$

Substituting this value for y in the first and third equations, we find

$$x = 2\cdot47, \qquad y = 3\cdot55, \qquad z = 1\cdot92.$$

The discussion of the P.E.'s of x, y, z will be found on page 107.

The quantity $[ll]$ evaluated above is not essential to the formation of the normal equations, but may be required later, in the discussion of P.E.'s.

Example 2. Given a set of 14 observational equations of the form

$$ax + by = l,$$

with weight p, the values of a, b, l, and p being given in the table below, form the normal equations.

a	b	l	s	p
0·47	1·38	−0·68	1·17	5
1·43	0·02	0·75	0·80	4
0·15	−1·06	0·60	−0·31	3
−0·98	−0·69	0·10	−1·57	3
−0·82	−1·20	1·05	−0·97	4
−0·88	−1·25	−1·00	−3·13	3
−1·58	−0·20	−0·32	−2·10	5
−1·12	0·62	−0·27	−0·77	3
−1·27	−0·17	1·47	0·03	3
−1·10	0·36	−0·06	−0·80	3
−1·14	1·19	−0·36	−0·31	5
−0·96	0·62	−0·36	−0·70	3
−0·13	1·58	0·02	1·47	5
0·62	1·00	0·50	2·12	3

We have to evaluate $[paa]$, $[pab]$, $[pal]$, etc.

The first step is to rewrite the above table, with each line multiplied by the corresponding weight.

pa	pb	pl	ps
2·35	6·90	−3·40	5·85
5·72	0·08	3·00	3·20
0·45	−3·18	1·80	−0·93
−2·94	−2·07	0·30	−4·71
−3·28	−4·80	4·20	−3·88
−2·64	−3·75	−3·00	−9·39
−7·90	−1·00	−1·60	−10·50
−3·36	1·86	−0·81	−2·31
−3·81	−0·51	4·41	0·09
−3·30	1·08	−0·18	−2·40
−5·70	5·95	−1·80	1·55
−2·88	1·86	−1·08	−2·10
−0·65	7·90	0·10	7·35
1·86	3·00	1·50	6·36

paa	pab	pal	pas
1·105	3·243	−1·598	2·750
8·180	·114	4·290	12·584
·068	−·477	·270	−·139
2·881	2·029	−·294	4·616
2·690	3·936	−3·444	3·182
2·323	3·300	2·640	8·263
12·482	1·580	2·528	16·590
3·763	−2·083	·907	2·587
4·839	·648	−5·601	−·115
3·630	−1·188	·198	2·640
6·498	−6·783	2·052	1·767
2·765	−1·786	1·037	2·016
·084	−1·027	−·013	−·955
1·153	1·860	·930	3·943
52·461	3·366	3·902	59·729

pab	pbb	pbl	pbs
	9·522	−4·692	8·073
	·002	·060	·176
	3·371	−1·905	·986
	1·428	−·207	3·250
	5·760	−5·040	4·656
	4·688	3·750	11·738
	·200	·320	2·100
	1·153	−·502	−1·432
	·087	−·750	−·016
	·389	−·065	−·864
	7·081	−2·142	−1·844
	1·153	−·670	−1·303
	12·482	·158	11·613
	3·000	1·500	6·360
3·366	50·316	−10·188	43·493

pal	*pbl*	*pll*	*pls*
		2·312	− 3·978
		2·250	6·600
		1·080	− ·558
		·030	− ·471
		4·410	− 4·074
		3·000	9·390
		·512	3·360
		·219	·624
		6·483	·133
		·011	·144
		·648	·558
		·389	·756
		·002	·147
		·750	3·180
3·902	− 10·188	22·096	15·811

The normal equations are therefore

$$52\cdot461x + 3\cdot366y - 3\cdot902 = 0,$$
$$3\cdot366x + 50\cdot316y + 10\cdot188 = 0,$$

and may easily be solved by ordinary algebraic methods.

The quantity $[pll]$ evaluated here; and $[ll]$ in the previous example, will be required in evaluating the P.E.'s of the unknowns. It is convenient, though not necessary, to evaluate $[ll]$ or $[pll]$ when forming the normal equations.

Example 3. A quantity l, of the form

$$x \cos\theta + y \sin\theta + z = l,$$

is observed for $\theta = 5°$, 15°, 25°, etc., up to $\theta = 355°$.

Form the normal equations.

With the previous notation, $a = \cos\theta$, $b = \sin\theta$, $c = 1$. Since

$$\cos^2\theta = \sin^2(90° - \theta),$$

$$[aa] = \overset{355°}{\underset{5°}{\Sigma}} \cos^2\theta = 2 \overset{175°}{\underset{5°}{\Sigma}} \cos^2\theta = 4 \overset{85°}{\underset{5°}{\Sigma}} \cos^2\theta = 18,$$

$$[bb] = \overset{355°}{\underset{5°}{\Sigma}} \sin^2\theta = 18,$$

$$[cc] = 36,$$
$$[ab] = \Sigma \cos\theta \sin\theta = 0,$$
$$[ac] = \Sigma \cos\theta = 0,$$
$$[bc] = \Sigma \sin\theta = 0.$$

Hence the normal equations are

$$18x = [al] = \Sigma l \cos\theta,$$
$$18y = [bl] = \Sigma l \sin\theta,$$
$$36z = [cl] = \Sigma l.$$

Example 4. The instants of passage of twelve successive swings of a pendulum are observed. Find the period.

Let the observed times of the twelve successive passages be t_0, t_1, t_2,...t_{11}. Let the true time of the first passage be a_0 and let the period be T. Then we have twelve observational equations,

$$\left.\begin{aligned} a_0 - t_0 \quad &= 0 \\ a_0 + T - t_1 &= 0 \\ a_0 + 2T - t_2 &= 0 \\ \cdots\cdots\cdots\cdots \\ \cdots\cdots\cdots\cdots \\ \cdots\cdots\cdots\cdots \\ a_0 + 11T - t_{11} &= 0 \end{aligned}\right\} \quad\ldots\ldots\ldots\ldots\ldots\ldots\ldots(1).$$

The normal equations are

$$12a_0 + (1+2+3+\ldots+11)\,T - (t_0 + t_1 + t_2 + \ldots + t_{11}) = 0,$$
$$(1+2+3+\ldots+11)\,a_0 + (1^2 + 2^2 + 3^2 + \ldots + 11^2)\,T - (t_1 + 2t_2 + \ldots + 11t_{11}) = 0.$$

If we substitute

$$t_0 + t_1 + t_2 + \ldots + t_{11} = s_1,$$
$$t_1 + 2t_2 + \ldots + 11t_{11} = s_2,$$

the normal equations become

$$12a_0 + \ 66T - s_1 = 0,$$
$$66a_0 + 506T - s_2 = 0.$$

Solving by the ordinary algebraic method, we find

$$T = \frac{2s_2 - 11s_1}{286}, \qquad a_0 = \frac{23s_1 - 3s_2}{78} \quad\ldots\ldots\ldots\ldots\ldots(2).$$

The following method of dealing with the above problem is entirely wrong, but affords an interesting example of the errors one is liable to commit.

Let the times be measured from the instant of the first passage, and let the times of the other passages be t_1, t_2,...t_{11}. Then the observational equations are

$$\left.\begin{aligned} t_1 - T \ &= 0 \\ t_2 - 2T &= 0 \\ t_3 - 3T &= 0 \\ \cdots\cdots\cdots \\ \cdots\cdots\cdots \\ \cdots\cdots\cdots \\ t_{11} - 11T &= 0 \end{aligned}\right\} \quad\ldots\ldots\ldots\ldots\ldots\ldots\ldots(1)'.$$

These equations are of equal weight, and the period is the value of T which makes

$$\Sigma\,(t_s - sT)^2 \text{ a minimum.}$$

Differentiating with respect to T, we find

$$T = \frac{t_1 + 2t_2 + \ldots + 11t_{11}}{1^2 + 2^2 + \ldots + 11^2} \quad\ldots\ldots\ldots\ldots\ldots\ldots(2)'.$$

Or, if we return to the notation originally used, where the time of the first passage is t_0,

$$T = \frac{s_2 - (1 + 2 + \ldots + 11)\,t_0}{1^2 + 2^2 + \ldots + 11^2}.$$

It should be noted that this result is entirely different from the one originally obtained in equation (2). This method of treatment is fallacious, because the error of noting the time of the first swing enters into each of the equations (1)′, $t_1, t_2, \ldots t_{11}$, being the differences of two observed times. The method of least squares is only applicable to observational equations which are independent, and it is therefore not permissible to apply it to equations (1)′. The only correct method of dealing with this problem is to write down equations (1), and solve them by the method followed above.

Example 5. Form the observational equations for the rate and error of a sidereal clock, from the following set of observations. Derive the normal equations, and solve them.

Star	Observed transit	R. A.
η Tauri	3$^\text{h}$ 38$^\text{m}$ 7$^\text{s}$·64	8$^\text{s}$·08
A Tauri	3 55 23 ·16	23 ·62
i Tauri	4 53 41 ·19	41 ·65
β Tauri	5 16 20 ·56	21 ·05

(The error of a clock at time t is $a + bt$ where a and b are constants.)

Example 6. From the equations

$$x \qquad\qquad = 2,$$
$$x + y \qquad\quad = 3,$$
$$x - y + z \quad = -1,$$
$$x - 3y + 2z = -2,$$

form the normal equations, and solve for x, y, z.

40. Solution of the Normal Equations.

In examples 2 and 5 above, there are only two unknowns in each case, so that there are only two normal equations; while in examples 1 and 6 there are three unknowns, but the coefficients of these unknowns in the normal equations are integral. In all four cases, therefore, the equations can be solved by the ordinary methods of elementary algebra. When the number of unknowns is greater than 2, and the coefficients in the normal equations are not integral, the solution has to be carried out by one of the special methods discussed in the next chapter.

One of the difficulties that arise when we attempt to solve the equations in the haphazard way of ordinary algebra is that of determining the number of places of decimals to retain at each step. It is one of the merits of Gauss's method (to be discussed in the next chapter), that it prevents our getting into difficulties over this.

CHAPTER VI

EVALUATION OF THE MOST PROBABLE VALUES OF THE
UNKNOWNS, THEIR WEIGHTS AND PROBABLE ERRORS

41. Gauss's Method of Substitution.

For the sake of simplicity in writing, we shall suppose that there are three unknowns, x, y, z, but the method can be automatically extended to include any number of unknowns.

Let the normal equations be

$$\left.\begin{aligned}
[aa]\,x + [ab]\,y + [ac]\,z - [al] &= 0 \\
+ [bb]\,y + [bc]\,z - [bl] &= 0 \\
+ [cc]\,z - [cl] &= 0
\end{aligned}\right\} \dots\dots\dots\dots(i).$$

From the first equation we find

$$x = -\frac{[ab]}{[aa]}\,y - \frac{[ac]}{[aa]}\,z + \frac{[al]}{[aa]} \dots\dots\dots\dots\dots(ii).$$

Substituting this value in the second and third equations, we obtain the following equations:

$$\left.\begin{aligned}
[bb\,1]\,y + [bc\,1]\,z - [bl\,1] &= 0 \\
+ [cc\,1]\,z - [cl\,1] &= 0
\end{aligned}\right\} \dots\dots\dots\dots(iii),$$

where

$$\left.\begin{aligned}
[bb\,1] &= [bb] - \frac{[ab]\,[ab]}{[aa]} \\
[bc\,1] &= [bc] - \frac{[ab]\,[ac]}{[aa]}
\end{aligned}\right\} \dots\dots\dots\dots(iv),$$

etc.

From the first equation in (iii),

$$y = -\frac{[bc\,1]}{[bb\,1]}\,z + \frac{[bl\,1]}{[bb\,1]} \dots\dots\dots\dots\dots(v).$$

Substituting this value in the second equation (iii), we find

$$[cc2]\, z - [cl2] = 0 \quad \dots\dots\dots\dots\dots\text{(vi)},$$

where
$$\left.\begin{aligned}[cc2] &= [cc1] - \frac{[bc1][bc1]}{[bb1]} \\[1em] [cl2] &= [cl1] - \frac{[bc1][bl1]}{[bb1]}\end{aligned}\right\} \dots\dots\dots\dots\text{(vii)}.$$

Equation (vi) gives the value of z. Substituting this value for z, in equation (v) we evaluate y, and then by substitution for y and z in equation (ii) we obtain the value of x. The equations from which the values of the unknowns are deduced are here collected for convenience of reference.

$$\left.\begin{aligned} z &= \frac{[cl2]}{[cc2]} \\[1em] y &= -\frac{[bc1]}{[bb1]}\, z + \frac{[bl1]}{[bb1]} \\[1em] x &= -\frac{[ab]}{[aa]}\, y - \frac{[ac]}{[aa]}\, z + \frac{[al]}{[aa]}\end{aligned}\right\} \dots\dots\dots\dots\text{(viii)}.$$

The solution for four unknowns is similar to the above, the work being carried through a further stage, involving the evaluation of $[dd3]$, etc. In practice, the order of procedure is slightly varied from that shown above, but the method can be best understood by the actual solution of a set of normal equations.

42. Checks in Computation.

As the solution of a system of normal equations involves a considerable amount of arithmetic, it is advisable to check the correctness of the work at each stage.

α. In the first place, it can be shown that the leading co-efficients $[aa]$, $[bb1]$, $[cc2]$, etc., are all positive. Since $[aa]$ is the sum of a number of squares, it must be positive. Again

$$\begin{aligned}[aa][bb1] &= [aa][bb] - [ab][ab] \\ &= \Sigma\,(a_r b_s - a_s b_r)^2 \\ &= \text{a positive quantity}^* \\ &= [KK], \text{ say.}\end{aligned}$$

Hence $[bb1]$ is positive.

* It follows that $[aa][bb]$ is greater than $[ab]^2$.

Similarly $[aa][cc1] = \Sigma\,(a_r c_s - a_s c_r)^2 = [LL]$, say.

With this notation,

$$[aa][bc1] = [aa][bc] - [ac][ab]$$
$$= \Sigma\,\{(a_r b_s - a_s b_r)(a_r c_s - a_s c_r)\}$$
$$= [KL].$$

It follows from equation (vii) that

$$[aa]^2[bb1][cc2] = [KK][LL] - [KL][KL]$$
$$= \Sigma\,(K_r L_s - K_s L_r)^2$$
$$= \text{a positive quantity.}$$

Hence it follows that $[cc2]$ is a positive quantity.

Since

$$[aa][bb1] = [KK], \quad [aa][bc1] = [KL], \quad \text{etc.,}$$

it follows that the system of equations obtained after eliminating x is similar to the original system of normal equations, in that all the coefficients in the leading diagonal are positive, while the other coefficients are symmetrical about this diagonal. The leading coefficient which results from eliminating two unknowns from this set of equations, will be of the same nature as $[cc2]$; i.e. it will be positive. This coefficient is $[dd3]$. This line of argument can be applied to any number of unknowns.

The calculation of $[bb1]$, $[cc2]$, etc., may in some measure be checked by the result deduced above, a negative value of such a coefficient indicating an error in computation.

β. A useful check on the calculations made in solving the normal equations, may be obtained by a simple extension of the checks on the formation of the normal equations (see § 37). The formation of the equations is checked by means of quantities $[as]$, $[bs]$, ..., $[ls]$. If we operate upon these quantities in the same way as upon $[al]$, $[bl]$, etc., we can form new quantities $[bs1]$, $[cs1]$, $[cs2]$, etc. When there are only three unknowns, it is easily shown that

$$[bb1] + [bc1] + [bl1] = [bs1],$$
$$[bc1] + [cc1] + [cl1] = [cs1].$$

After the next stage of elimination we have

$$[cc2] + [cl2] = [cs2].$$

Similar relations will hold for any number of unknowns. A check of this kind is in general sufficient to detect any errors of computation. The only additional work involved in the carrying on of the check is the addition of a column to the tabular form in which the work is carried out.

γ. An additional check may be got by reversing the order of elimination of the unknowns, but this involves considerable labour.

δ. When the unknowns have been evaluated, the substitution of their values in the observational equations yields the values of the residuals v_1, v_2, etc. A number of useful checks on the work can be derived from these residuals. It has already been shown that

$$[av] = [bv] = [cv] = \ldots = 0.$$

Any of these equations may be used as a check on the final results of solving the normal equations.

ε. If the observational equations, written in the form

$$a_r x + b_r y + c_r z - l_r = v_r \qquad (r = 1, 2, 3, \text{etc.})$$

be multiplied by v_1, v_2, v_3, etc., and added, we obtain the equation

$$-[vl] = [vv],$$

since the coefficients of x, y, z, etc., vanish.

If the observational equations be multiplied by l_1, l_2, etc., and added, we obtain the equation

$$[al]\,x + [bl]\,y + [cl]\,z - [ll] = [lv] = -[vv]\ldots\ldots\ldots(\text{ix}).$$

Now take equations (viii) and substitute in the last equation (ix) for x, y, z, in turn. Substituting for x, we obtain

$$-[bl\,1]\,y - [cl\,1]\,z + [ll\,1] = [vv].$$

Substituting for y, from the second of equations (viii),

$$-[cl\,2]\,z + [ll\,2] = [vv].$$

Finally substituting for z, from the first of equations (viii),

$$[ll\,3] = [vv].$$

Hence
$$[vv] = [ll\,3] = [ll\,2] - \frac{[cl\,2]^2}{[cc\,2]}$$

or
$$[vv] = [ll] - \frac{[al]^2}{[aa]} - \frac{[bl\,1]^2}{[bb\,1]} - \frac{[cl\,2]^2}{[cc\,2]}.$$

Any of the equations given above might be used as a check on the final result. Perhaps the most useful check is to compare the value of $[vv]$ obtained by squaring the calculated residuals, with the value of $[vv]$ obtained from the last equation. In any case the value of $[vv]$ will be required when we come to the problem of evaluating the probable errors of the adjusted values of the unknowns.

It should be noted that the last form for $[vv]$ is capable of immediate generalisation for any number of unknowns, say m. We then have

$$[vv] = [llm] = [ll] - \frac{[al]^2}{[aa]} - \frac{[bl1]^2}{[bb1]} - \text{etc.,}$$

where there are m terms after $[ll]$ on the right-hand side of the equation.

43. Form of Solution.

In any piece of work it is a decided advantage to follow a systematic method of solution, in which the evaluation of co-efficients and the application of checks is to a large extent mechanical. Table A shows a convenient form of solution for four unknowns.

When there are only three unknowns, there will be no terms involving d, and only the first eleven lines in the table need be evaluated. When an arithmometer, or a table of products, is used, the solution can be worked out line by line as shown in the table. The form can be varied to suit individual tastes. The last column indicates how the various lines in the table are derived from the preceding lines. (Ch.) indicates the lines in which the checks are to be applied. When the work is to be done by the use of logarithmic tables, the order of computation need not be altered, but it is necessary to put in some additional lines containing logarithms. We shall solve by the use of logarithmic tables, following the order of Table A as closely as possible, the following set of normal equations:

$$
\begin{aligned}
153 \cdot 000x + 6 \cdot 285y + 2 \cdot 485z - 27 \cdot 831w &= 22 \cdot 093 \\
+ 8 \cdot 989y + 4 \cdot 037z - 0 \cdot 426w &= -3 \cdot 855 \\
+ 23 \cdot 616z - 3 \cdot 504w &= -9 \cdot 952 \\
+ 9 \cdot 080w &= 7 \cdot 251
\end{aligned}
$$

Check
156·032
15·030
16·682
−15·430

TABLE A.

#	[aa]	[ab]	[ac]	[ad]	[al]	[as]	Operation
1	$[aa]$	$[ab]$	$[ac]$	$[ad]$	$[al]$	$[as]$	
2	$x=\dfrac{[al]}{[aa]}$	$\dfrac{[ab]}{[aa]}$	$\dfrac{[ac]}{[aa]}$	$\dfrac{[ad]}{[aa]}$	$\dfrac{[al]}{[aa]}$	$\dfrac{[as]}{[aa]}$	$1 \div [aa]$
3	$-\dfrac{[ad]}{[aa]}\,w$	$[bb]$	$[bc]$	$[bd]$	$[bl]$	$[bs]$	
4	$-\dfrac{[ac]}{[aa]}\,z$	$\dfrac{[ab]}{[aa]}[ab]$	$\dfrac{[ab]}{[aa]}[ac]$	$\dfrac{[ab]}{[aa]}[ad]$	$\dfrac{[ab]}{[aa]}[al]$	$\dfrac{[ab]}{[aa]}[as]$	$2 \times [ab]$
5	$-\dfrac{[ab]}{[aa]}\,y.$	$[bb1]$	$[bc1]$	$[bd1]$	$[bl1]$	$[bs1]$	$(3-4)$ Ch.
6	⋮	$y=\dfrac{[bl1]}{[bb1]}$	$\dfrac{[bc1]}{[bb1]}$	$\dfrac{[bd1]}{[bb1]}$	$\dfrac{[bl1]}{[bb1]}$	$\dfrac{[bs1]}{[bb1]}$	$5 \div [bb1]$
7	⋮	$-\dfrac{[bd1]}{[bb1]}\,w$	$[cc]$	$[cd]$	$[cl]$	$[cs]$	
8	⋮	$-\dfrac{[bc1]}{[bb1]}\,z,$	$\dfrac{[ac]}{[aa]}[ac]$	$\dfrac{[ac]}{[aa]}[ad]$	$\dfrac{[ac]}{[aa]}[al]$	$\dfrac{[ac]}{[aa]}[as]$	$2 \times [ac]$
9	⋮	⋮	$[cc1]$	$[cd1]$	$[cl1]$	$[cs1]$	$(7-8)$ Ch.
10	⋮	⋮	$\dfrac{[bc1]}{[bb1]}[bc1]$	$\dfrac{[bc1]}{[bb1]}[bd1]$	$\dfrac{[bc1]}{[bb1]}[bl1]$	$\dfrac{[bc1]}{[bb1]}[bs1]$	$6 \times [bc1]$
11	⋮	⋮	$[cc2]$	$[cd2]$	$[cl2]$	$[cs2]$	$(9-10)$ Ch.
12	⋮	⋮	$z=\dfrac{[cl2]}{[cc2]}$	$\dfrac{[cd2]}{[cc2]}$	$\dfrac{[cl2]}{[cc2]}$	$\dfrac{[cs2]}{[cc2]}$	$11 \div [cc2]$
13	⋮	⋮	$-\dfrac{[cd2]}{[cc2]}\,w.$	$[dd]$	$[dl]$	$[ds]$	
14	⋮	⋮	⋮	$\dfrac{[ad]}{[aa]}[ad]$	$\dfrac{[ad]}{[aa]}[al]$	$\dfrac{[ad]}{[aa]}[as]$	$2 \times [ad]$
15	⋮	⋮	⋮	$\dfrac{[bd1]}{[bb1]}[bd1]$	$\dfrac{[bd1]}{[bb1]}[bl1]$	$\dfrac{[bd1]}{[bb1]}[bs1]$	$6 \times [bd1]$
16	⋮	⋮	⋮	$\dfrac{[cd2]}{[cc2]}[cd2]$	$\dfrac{[cd2]}{[cc2]}[cl2]$	$\dfrac{[cd2]}{[cc2]}[cs2]$	$12 \times [cd2]$
17	⋮	⋮	⋮	$[dd3]$	$[dl3]$	$[ds3]$	$(13{-}14{-}15{-}16)$ Ch.

$$w=\frac{[dl3]}{[dd3]}$$

The solution of this set of equations is given in Table B. In this table the order of solution is that indicated in Table A. The last column indicates a comparison between this table and Table A, and the additional lines necessitated by the use of logarithmic tables are indicated by accented numbers in the first column.

A few points of procedure deserve notice. Line 1′ contains the logarithms of the quantities in line 1. The logarithm of a negative quantity is expressed by writing down the logarithm of the quantity concerned, with n prefixed, to indicate that we are dealing with a negative quantity. Thus

$$\log - 27{\cdot}831 = n\, 1{\cdot}4445288,$$

$1{\cdot}4445288$ being $\log 27{\cdot}831$. Two n's added or subtracted annihilate one another. Line 2 is derived from line 1′ by subtracting the first term in line 1′ from all the others. This can be conveniently done by writing the first term at the bottom of a piece of paper, and carrying the piece of paper along line 1′, so that this term may be subtracted in turn from each of the other terms. To derive line 4 from line 2, we first add $0{\cdot}7983053$ to each term in line 2, and find from the tables the numbers corresponding to the sums. These numbers are written down in line 4. The remainder of the work follows the same method.

The final solution is

$$x = 0{\cdot}727,$$
$$y = -0{\cdot}836,$$
$$z = 0{\cdot}093,$$
$$w = 3{\cdot}026.$$

44. The Doolittle Method of Solution.

A variant of Gauss's method of substitution, due to Mr Doolittle, of the U.S. Coast and Geodetic Survey*, yields a considerable gain in speed, and saves much labour by reducing the number of entries from the tables to a minimum. Take the case of three normal equations

$$[aa]\,x + [ab]\,y + [ac]\,z = [al]$$
$$+\,[bb]\,y + [bc]\,z = [bl]$$
$$+\,[cc]\,z = [cl].$$

* *Coast and Geodetic Survey, Report for* 1878, Appendix 8, pp. 115–118.

TABLE B.

	x	y	z	w		Check	Remarks
1	153·000	6·285	2·485	−27·831	22·093	156·032	log 1 — A1
1'	2·1846914	0·7983053	0·3953264	n1·4445288	1·3442547	2·1932137	1' − 2·1846914 log — A2
2	x = [1̄·1595633]	2̄·6136169	2̄·2106350	n1̄·2598374	1̄·1695633	0·0085223	numbers — A3
3	+[1̄·2598374]w	8·989	4·037	−0·426	−3·855	15·030	,, — A4
4	−[2̄·2106350]z	0·258	0·102	−1·143	0·908	6·410	,, 3—4 — A5
5	−[2̄·6136169]y	8·731	3·935	0·717	−4·763	8·621	log 5
5'	= ·14440	0·9410640	0·5949447	1̄·8555192	n0·6778806	0·9355073	5' − 0·9410640 log — A6
6	+·54985	y = [n1̄·7368166]	1̄·6538807	2̄·9144552	n1̄·7368166	1̄·994433	numbers — A7
7	−·00152	−[2̄·9144552]w	23·616	−3·504	−9·952	16·682	,, — A8
8	+·03434	−[1̄·6658807]z	0·040	−0·452	0·359	2·534	,, 7—8 — A9
9	x = 0·727.	= −·5455	23·576	−3·052	−10·311	14·148	,, — A10
10	⋮	−·2482	1·773	0·323	−2·147	3·885	,, 9—10 — A11
11	⋮	−·0421	21·803	−3·375	−8·164	10·263	log 11
11'	⋮	y = −·836.	1·3385163	n0·5288738	n0·9119030	1·0112743	11' − 1·3385163 log — A12
12	⋮	⋮	z = [n1·5733867]	n1̄·1897575	n1̄·5733867	1̄·6727580	numbers — A13
13	⋮	⋮	−[n1̄·1897575]w	9·080	7·251	−15·430	,, — A14
14	⋮	⋮	= ·4679	5·063	−4·019	−28·383	,, 13—14 — A15
14'	⋮	⋮	−·3744	4·017	11·270	12·953	,, 14'—15 — A16
15	⋮	⋮	z = 0·093.	0·059	−0·391	0·708	,, 15'—16 — A17
15'	⋮	⋮	⋮	3·958	11·661	12·245	
16	⋮	⋮	⋮	0·522	1·264	−1·589	
17	⋮	⋮	⋮	3·436	10·397	13·834	
				0·5365032	1·0169080		
				w = 3·026	log w = 0·480048		

The notation used in the solution is that of Gauss's method. The order of solution is shown in tables C and D.

TABLE C.

	x	y	z	
$-\dfrac{1}{[aa]}$	$[aa]$ $x=$	$[ab]$ $-\dfrac{[ab]}{[aa]}$	$[ac]$ $-\dfrac{[ac]}{[aa]}$	$-[al]$ $+\dfrac{[al]}{[aa]}$
$-\dfrac{1}{[bb1]}$		$[bb1]$ $y=$	$[bc1]$ $-\dfrac{[bc1]}{[bb1]}$	$[bl1]$ $+\dfrac{[bl1]}{[bb1]}$
$-\dfrac{1}{[cc2]}$			$[cc2]$ $z=$	$-[cl2]$ $\dfrac{[cl2]}{[cc2]}$

TABLE D.

y	z	
$[bb]$ $-[ab]\dfrac{[ab]}{[aa]}$	$[bc]$ $-[ab]\dfrac{[ac]}{[aa]}$	$-[bl]$ $[ab]\dfrac{[al]}{[aa]}$
	$[cc]$ $-[ac]\dfrac{[ac]}{[aa]}$ $-[bc1]\dfrac{[bc1]}{[bb1]}$	$-[cl]$ $[ac]\dfrac{[al]}{[aa]}$ $[bc1]\dfrac{[bl1]}{[bb1]}$

The coefficients of the first normal equation are written in line 1, table C. The reciprocal of $[aa]$, with its sign changed, is written in the first column, and all the other quantities in line 1 are multiplied by it, the results being entered in line 2. Line 2 then gives x as an explicit function of y and z. The coefficients of the second normal equation are written in line 1, table D. Line 2, table C, is now multiplied by $[ab]$, and the results entered in table D, line 2. The sum of lines 1 and 2, table D, is entered in table C, line 3. The reciprocal of $[bb1]$ is written in the first column of table C, and all the other quantities in line 3 are multiplied by this reciprocal, the results being entered in line 4. This line gives y as an explicit function of z. Next, the coefficients of the third normal equation are entered in line 3 of table D. The terms in lines 2 and 4 of table C, beginning only at the z terms, are multiplied by $[ac]$ and $[bc1]$ respectively, and the results entered in table D, lines 4 and 5. The sum of lines 3, 4 and 5 of table D is entered in table C, line 5. Line 6, table C, is obtained by first writing down the reciprocal of $[cc2]$, with its sign changed, and multiplying the other term of line 5 by this residual. Line 6 gives the value of z. The values of y and x are obtained by successive substitution in lines 4 and 2 of table C.

This method of solution can be easily extended for any number of unknowns. Its advantage over the older method, due to the reduction of entries, is even greater for larger numbers of unknowns than for three unknowns as discussed above. In practice it is convenient to make tables C and D on separate sheets of paper, and to fold table C along alternate lines, to facilitate the carrying of the numbers of table C to where they are required in table D.

45. Solution by the Method of Determinants*.

The solution of the normal equations

$$[aa]\,x + [ab]\,y + [ac]\,z = [al]$$
$$+ [bb]\,y + [bc]\,z = [bl]$$
$$+ [cc]\,z = [cl]$$

may be immediately written down in determinantal form

$$x = \frac{1}{D}\begin{vmatrix} [al] & [ab] & [ac] \\ [bl] & [bb] & [bc] \\ [cl] & [bc] & [cc] \end{vmatrix}, \qquad y = \frac{1}{D}\begin{vmatrix} [aa] & [al] & [ac] \\ [ab] & [bl] & [bc] \\ [ac] & [cl] & [cc] \end{vmatrix},$$

etc.,

where

$$D = \begin{vmatrix} [aa] & [ab] & [ac] \\ [ab] & [bb] & [bc] \\ [ac] & [bc] & [cc] \end{vmatrix}.$$

---

* This method is not a very practical one in general. It is given here because it is useful in the development of the theory.

These results may be extended to any number of unknowns. The determinant D, which is symmetrical about the leading diagonal, contains as many rows and columns as there are unknowns.

The above solution leads to a form for calculating $[vv]$ which is occasionally useful. It has already been shown (page 93) that

$$[vv] = [ll] - [al]\,x - [bl]\,y - [cl]\,z.$$

$$\therefore \quad D\,[vv] = [ll] \times \begin{vmatrix} [aa] & [ab] & [ac] \\ [ab] & [bb] & [bc] \\ [ac] & [bc] & [cc] \end{vmatrix} + [al] \begin{vmatrix} [ab] & [ac] & [al] \\ [bb] & [bc] & [bl] \\ [bc] & [cc] & [cl] \end{vmatrix}$$

$$- [bl] \begin{vmatrix} [aa] & [ac] & [al] \\ [ab] & [bc] & [bl] \\ [ac] & [cc] & [cl] \end{vmatrix} + [cl] \begin{vmatrix} [aa] & [ab] & [al] \\ [ab] & [bb] & [bl] \\ [ac] & [bc] & [cl] \end{vmatrix}.$$

$$\therefore \quad D\,[vv] = \begin{vmatrix} [aa] & [ab] & [ac] & [al] \\ [ab] & [bb] & [bc] & [bl] \\ [ac] & [bc] & [cc] & [cl] \\ [al] & [bl] & [cl] & [ll] \end{vmatrix}.$$

This is a determinant formed from D by the addition of a fresh row and a fresh column containing the l terms.

46. Probable Error of an Observational Equation of Unit Weight, when there are n Observations involving m Independent Unknowns.

If the observations be not all of the same weight, each observational equation is first multiplied by the square root of its weight, so as to reduce all the residuals to unit weight. Let the system of residuals, thus reduced to unit weight, be v_1, v_2, etc. If the error law followed by the residuals be

$$\frac{h}{\sqrt{\pi}}\, e^{-h^2 \Delta^2},$$

then the *a priori* probability of the given system of residuals v_1, v_2, etc., is

$$C h^n e^{-h^2 [vv]},$$

where C is a numerical constant (cf. § 14). This probability may be written

$$C h^n e^{-h^2 \Sigma\,(a_r x + b_r y + c_r z + \dots - l_r)^2}.$$

But x, y, etc., are all unknown and independent, and may, as far as we know, take any values between $-\infty$ and $+\infty$. The complete probability is therefore got by integrating the last expression between limits $\pm\infty$ for each of the unknowns. The probability then becomes

$$C\int_{-\infty}^{+\infty}\int_{-\infty}^{+\infty}\ldots\ldots\int_{-\infty}^{+\infty} h^n e^{-h^2\Sigma\,(a_r x + b_r y + \ldots - l_r)^2}\,dx\,dy\,dz\ldots.$$

$$m \text{ Integrations}$$

In order to integrate with respect to x, we write the coefficient of $-h^2$ in the exponential term, in the form

$$P + (Qx + R)^2,$$

where Q is a numerical factor, and P and R are functions of y, z, etc.

$$\int_{-\infty}^{+\infty} h e^{-h^2\{P+(Qx+R)^2\}}\,dx = h e^{-h^2 P}\int_{-\infty}^{+\infty} e^{-h^2(Qx+R)^2}\,dx$$

$$= \frac{\sqrt{\pi}}{Q}\cdot e^{-h^2 P}.$$

The factor $\dfrac{\sqrt{\pi}}{Q}$ is a numerical factor which depends only on the values of the coefficients in the observational equations. We may regard it as being absorbed in the constant C.

The expression for the probability, with x eliminated, becomes

$$C\int_{-\infty}^{+\infty}\int_{-\infty}^{+\infty}\ldots\ldots\int_{-\infty}^{+\infty} h^{n-1} e^{-h^2 P}\,dy\,dz\ldots.$$

$$(m-1) \text{ Integrations}$$

Here P is the quadratic function of y, z, etc., which results from

$$\Sigma\,(a_r x + b_r y + \ldots - l_r)^2,$$

when x is so chosen as to make this last expression a minimum. It follows that after m integrations, the probability becomes

$$Ch^{n-m} e^{-h^2 T},$$

where C is a constant, and T is the value of $\Sigma(a_r x + b_r y + \ldots - l_r)^2$ which results from giving x, y, z, etc., the values which make this sum a minimum. But these are the values of x, y, z, etc., which are determined by the ordinary least square solution, using the normal equations. It follows that

$$T = [vv].$$

The total probability of the given system of residuals is thus

$$Ch^{n-m}e^{-h^2[vv]}.$$

The value of the parameter h which determines the scale of the error curve has to be selected so as to make the probability of the given system of residuals a maximum. Taking logarithms, and differentiating with respect to h, we find

$$\frac{n-m}{h} - 2h\,[vv] = 0,$$

or

$$h = \sqrt{\frac{n-m}{2\,[vv]}}.$$

The M.S.E. μ of an observation of unit weight is thus

$$\mu = \sqrt{\frac{[vv]}{n-m}},$$

and the P.E. r of an observation of unit weight, or an observational equation of unit weight, is given by*

$$r = 0{\cdot}6745\,\sqrt{\frac{[vv]}{n-m}}.$$

The quantity $[vv]$ may be evaluated in a number of ways. The obvious method is to calculate it from the actual residuals, and since it is generally advisable to test the character of the work by evaluating the residuals and considering their magnitude, this is perhaps the most satisfactory method. But it is also possible to calculate $[vv]$ from any of the relations shown below.

$$[vv] = -\,[vl],$$
$$[vv] = [ll] - [al]\,x - [bl]\,y - [cl]\,z - \text{etc.},$$
$$[vv] = [ll] - \frac{[al]^2}{[aa]} - \frac{[bl1]^2}{[bb1]} - \frac{[cl2]^2}{[cc2]} - \text{etc.}$$

Or any one of these relations may be used as a check upon the value obtained by squaring the residuals.

---

* Or $r = 0{\cdot}6745\,\sqrt{\frac{[pvv]}{n-m}}$, where v is the residual in an observational equation of weight p.

47. Evaluation of Probable Errors of the Unknowns.

Since the normal equations are linear in x, y, z, etc., and also in l_1, l_2, etc., we may write

$$x = \alpha_1 l_1 + \alpha_2 l_2 + \text{etc.,}$$
$$y = \beta_1 l_1 + \beta_2 l_2 + \text{etc.,}$$
$$\dots\dots\dots\dots\dots\dots$$
$$\dots\dots\dots\dots\dots\dots$$

where α_1, α_2, ..., β_1, β_2, ..., etc., are functions of the coefficients a, b, c.

The P.E. of an observational equation of unit weight, or the P.E. of each of the l's, is the quantity r which has already been deduced above. If r_x, r_y, etc., be the P.E.'s of the unknowns, it follows that

$$r_x{}^2 = r^2 [\alpha\alpha], \quad r_y{}^2 = r^2 [\beta\beta], \quad \text{etc.}$$

It is therefore necessary to evaluate $[\alpha\alpha]$, $[\beta\beta]$, etc. The reciprocals of these sums may be regarded as the weights of x, y, etc. Calling these p_x, p_y, etc., we have the relations

$$p_x = \frac{1}{[\alpha\alpha]}, \qquad p_x r_x{}^2 = r^2.$$

Returning to the solution of the normal equations by determinants, we may write

$$x = \frac{1}{D} \begin{vmatrix} [al] & [ab] & [ac] \\ [bl] & [bb] & [bc] \\ [cl] & [bc] & [cc] \end{vmatrix} \equiv \alpha_1 l_1 + \alpha_2 l_2 + \dots .$$

Collecting the terms in l_1 in the determinant, we have the relation

$$D\alpha_1 = \begin{vmatrix} a_1 & [ab] & [ac] \\ b_1 & [bb] & [bc] \\ c_1 & [bc] & [cc] \end{vmatrix} = a_1 A + b_1 B + c_1 C,$$

where A, B, C are the minors of a_1, b_1, c_1, in the determinant. Squaring this expression, we may write the result as

$$D^2\alpha_1{}^2 = A \, (a_1 a_1 A + a_1 b_1 B + a_1 c_1 C)$$
$$+ B \, (a_1 b_1 A + b_1 b_1 B + b_1 c_1 C)$$
$$+ C \, (a_1 c_1 A + b_1 c_1 B + c_1 c_1 C).$$

Summing for all possible suffixes we find

$$D^2[\alpha\alpha] = A\,\{[aa]\,A + [ab]\,B + [ac]\,C\}$$
$$+ B\,\{[ab]\,A + [bb]\,B + [bc]\,C\}$$
$$+ C\,\{[ac]\,A + [bc]\,B + [cc]\,C\}$$
$$= A\{[aa]\,A + [ab]\,B + [ac]\,C\} = A\,D.$$

For if the second and third brackets be written in the form of determinants, two columns of each determinant would be the same, and so their values would be zero.

Hence
$$[\alpha\alpha] = \frac{A}{D} = \begin{vmatrix} 1 & [ab] & [ac] \\ 0 & [bb] & [bc] \\ 0 & [bc] & [cc] \end{vmatrix} \div D.$$

Comparing this with the value of x deduced above, we see that $[\alpha\alpha]$ is the value of x which we should deduce if we put

$$[al] = 1, \quad [bl] = [cl] = 0.$$

Similarly $[\beta\beta]$ is the value of y obtained when we put

$$[al] = [cl] = 0 \quad \text{and} \quad [bl] = 1.$$

Again
$$D \cdot \beta = \begin{vmatrix} [aa] & a_1 & [ac] \\ [ab] & b_1 & [bc] \\ [ac] & c_1 & [cc] \end{vmatrix} = a_1 B + b_1 F + c_1 G,$$

where B, F, and G are the minors of a_1, b_1, c_1, in the determinant. B is the same as the B given above.

We thus have
$$D\alpha = a_1 A + b_1 B + c_1 C.$$
$$D\beta = a_1 B + b_1 F + c_1 G.$$
$$\therefore \quad D^2\alpha\beta = A\,\{a_1 a_1 B + a_1 b_1 F + a_1 c_1 G\}$$
$$+ B\,\{a_1 b_1 B + b_1 b_1 F + b_1 c_1 G\}$$
$$+ C\,\{a_1 c_1 B + b_1 c_1 F + c_1 c_1 G\}$$
$$D^2[\alpha\beta] = A\,\{[aa]\,B + [ab]\,F + [ac]\,G\}$$
$$+ B\,\{[ab]\,B + [bb]\,F + [bc]\,G\}$$
$$+ C\,\{[ac]\,B + [bc]\,F + [cc]\,G\}$$
$$= D \cdot B,$$

the first and last brackets being zero, as can easily be shown by writing them in determinant form. It follows that

$$D\,[\alpha\beta] = B = \begin{vmatrix} [aa] & 1 & [ac] \\ [ab] & 0 & [bc] \\ [ac] & 0 & [cc] \end{vmatrix} = \begin{vmatrix} 0 & [ab] & [ac] \\ 1 & [bb] & [bc] \\ 0 & [bc] & [cc] \end{vmatrix},$$

$[\alpha\beta]$ is therefore the value of y obtained when

$$[al] = 1, \quad [bl] = [cl] = 0,$$

or the value of x obtained when

$$[al] = [cl] = 0, \quad [bl] = 1.$$

48. Evaluation of the Weights of x, y, z, etc.

The weight of the unknown which is first evaluated in the Gauss solution can be immediately deduced. If in the equations of § 41, we put

$$[al] = [bl] = 0 \text{ and } [cl] = 1,$$

then $\qquad\qquad [bl1] = 0, \text{ and } [cl2] = 1.$

Equation (vi) then yields $[cc2][\gamma\gamma] = 1,$

or $\qquad\qquad p_z = \dfrac{1}{[\gamma\gamma]} = [cc2].$

The quantity $[cc2]$ is one of the quantities evaluated in the process of solving the normal equations. In the general case where there are m unknowns, the weight of the last unknown in the order of elimination is the coefficient of that unknown in the last elimination equation. Thus in table A, page 95, the weight of w is given by

$$p_w = [dd3].$$

The weight of any unknown might be evaluated by making that particular unknown the last in the order of elimination. If the normal equations involve a large number of unknowns, of which it is only necessary to find the weights of a few, it is a considerable advantage to eliminate last the unknowns whose weights are required.

The following special cases should be particularly noted.

(a) *When there are only two unknowns, x and y.*

With the usual notation

$$p_y = [bb1] = \frac{[aa].[bb] - [ab].[ab]}{[aa]}.$$

And similarly, $$p_x = \frac{[aa][bb] - [ab][ab]}{[bb]}.$$

(b) *When there are three unknowns, x, y, z.*

It has been shown in § 47, that

$$p_x = \frac{1}{[aa]} = \frac{D}{A} = \frac{D}{[bb][cc] - [bc][bc]}.$$

Similarly

$$p_y = \frac{D}{[cc][aa] - [ac][ac]}, \qquad p_z = \frac{D}{[aa][bb] - [ab][ab]},$$

where D is the determinant formed by the coefficients of the normal equations. It may also be written

$$D = [aa][bb][cc] + 2[ab][bc][ca] - [aa][bc]^2 - [bb][ca]^2 - [cc][ab]^2.$$

When the normal equations contain only integral coefficients, it is generally simpler to solve them by the methods of ordinary algebra, and to calculate the weights by means of the above equations.

(c) *For any number of unknowns, it is easy to write down the weight of the last but one of the unknowns.*

Take the case of four unknowns eliminated in the order x, y, z, w. Then the weight of w is $[dd3]$. If we re-modelled our solution, and eliminated in the order x, y, w, z, the auxiliary quantities evaluated in the solution would remain unaltered as far as affix 2. The normal equations remaining after eliminating x and y are

$$[dd2]\,w + [cd2]\,z - [dl2] = 0,$$

$$[cd2]\,w + [cc2]\,z - [cl2] = 0.$$

From the first of these,

$$w = -\frac{[cd2]}{[dd2]}\,z + \frac{[dl2]}{[dd2]}.$$

Substituting this in the last equation, we find

$$p_z = \text{coefficient of } z \text{ in the final equation}$$

$$= [cc\,2] - [cd\,2]\frac{[cd\,2]}{[dd\,2]}$$

$$= \frac{[cc\,2]}{[dd\,2]}\left\{[dd\,2] - \frac{[cd\,2][cd\,2]}{[cc\,2]}\right\}.$$

$$p_z = \frac{[cc\,2][dd\,3]}{[dd\,2]}.$$

Thus p_z is easily determined. For $[cc\,2]$ is the last coefficient in the $[cc]$ column, $[dd\,3]$ the last coefficient in the $[dd]$ column, and $[dd\,2]$ the last coefficient but one in the $[dd]$ column.

Thus in the example worked in table B, page 97, $p_w = 3\cdot436$, and

$$p_z = \frac{21\cdot803 \times 3\cdot436}{3\cdot958}.$$

EXAMPLES.

1. Continuation of Ex. 1, page 82.

To find the weights and P.E.'s of x, y, z. Using the formulae of case (b) above, we find

$$D = 27 \times 15 \times 54 - 27 - 54 \times 6^2 = 19899,$$

$$p_x = \frac{19899}{15 \times 54 - 1} = 24\cdot6,$$

$$p_y = \frac{19899}{27 \times 54} = 13\cdot7,$$

$$p_z = \frac{19899}{27 \times 15 - 36} = 54\cdot0.$$

Substituting the derived values of the unknowns in the observational equations, we obtain the following residuals,

$$+\cdot24 \quad +\cdot09 \quad -\cdot11 \quad -\cdot06.$$

The P.E. of a single observational equation is

$$0\cdot6745 \sqrt{(\cdot24)^2 + (\cdot09)^2 + (\cdot11)^2 + (\cdot06)^2},$$

or   0·19.

$$\therefore \; r_x = \frac{0\cdot19}{\sqrt{24\cdot6}} = 0\cdot038,$$

$$r_y = \frac{0\cdot19}{\sqrt{13\cdot7}} = 0\cdot052,$$

$$r_z = \frac{0\cdot19}{\sqrt{54\cdot0}} = 0\cdot026.$$

2. Solve the normal equations of Ex. 2, page 84, evaluate the weights of the unknowns x and y, and from the formula

$$[pvv] = [pll] - \frac{[pal]^2}{[paa]} - \frac{[pbl1]^2}{[pbb1]}$$

find the P.E. of an observational equation of unit weight, and the P.E.'s of x and y.

3. Verify the result of the last example by evaluating $[pvv]$ from the residuals.

4. Prove that in the case of four unknowns x, y, z, w, eliminated in this order, the weight of y is

$$p_y = \frac{[bb1][cc2][dd3]}{[cc1][dd2]_b},$$

where all the auxiliaries have the usual meaning except $[dd2]_b$, which is the value of $[dd2]$ when the unknowns are eliminated in the order x, z, w, y.

The product $[aa][bb1][cc2][dd3]$ is the common denominator of x, y, z, w, when the expressions for the unknowns are reduced to the same denominator. The value of the product is therefore independent of the order of elimination.

$$\therefore \quad [aa][bb1][cc2][dd3] = [aa][cc1][dd2]_b[bb3],$$

whence the result immediately follows.

Deduce from table B the weight of y.

5. Show for the case of three unknowns, that the weights of the unknowns must always be positive.

6. In Ex. 3, page 86, find the weights of the unknowns.

7. In the same example show that

$$[vv] = [ll] - 36z^2 - 18\,(x^2 + y^2).$$

Write down the equations of condition in the form

$$v_r + l_r = x \cos\theta + y \sin\theta + z.$$

Multiply by l_r and add, remembering the relation

$$[vv] = -[vl].$$

49. An Alternative Method of finding the Weights of the Unknowns.

It has been shown that the weight of x is the value of x derived by solving a new set of normal equations in which

$$[al] = 1, \quad [bl] = [cl] = \ldots = 0.$$

The weight of each unknown can in fact be derived by solving the appropriate set of normal equations. In practice the weights can be conveniently found by combining the solution of all these subsidiary sets of normal equations with the ordinary solution for the unknowns. For the case of three unknowns, three additional columns are added to the table of solution, with 1, 0, 0 ; 0, 1, 0 ; and 0, 0, 1; written in positions corresponding to $[al]$, $[bl]$ and $[cl]$. If these columns be headed R, S and T respectively, the values of z derived in these columns are

$$[\alpha\gamma], \quad [\beta\gamma], \quad [\gamma\gamma].$$

The derivation of $[\alpha\alpha]$ and $[\beta\beta]$ follows quite simply from these. The method can be most clearly understood from the working of an actual example.

The following set of normal equations were derived by H. N. Russell in some work on stellar parallax :

$$8\cdot000x + 4\cdot106y + 0\cdot698z = -2588$$
$$+ 3\cdot989y + 0\cdot051z = -1441$$
$$+ 4\cdot613z = 1349.$$

The work of solution is considerably simplified by changing the unit of the absolute terms to 1000 times its original value. For, as the equations stand, the check sum for the first equation is $2565\cdot196$; while in the equation obtained by our suggested change of unit, the check sum is $10\cdot216$, containing five significant figures instead of seven.

In writing the absolute terms in the table of solution, we change the unit, so that they become $- 2\cdot588, -1\cdot441, 1\cdot349$.

The solution is given on page 110. The last column indicates a comparison with the order of solution in table A, page 95. The last column but one shows how each line is derived from those which precede it. The first three lines in the check column give $[as]+1$, $[bs]+1$, $[cs]+1$.

The work of solution was carried out by means of a Brunsviga calculator. Division of a whole line by any quantity, as for example, the division of line 8 by $[bb1]$, i.e. by $1\cdot8826$, was carried out by multiplying the whole line by the reciprocal of $[bb1]$.

The last line of the table gives the values of z, $[\alpha\gamma]$, $[\beta\gamma]$, $[\gamma\gamma]$.

$z = 0\cdot3453$, $[\alpha\gamma] = -0\cdot0357$, $[\beta\gamma] = 0\cdot0362$, $[\gamma\gamma] = 0\cdot2221$.

From line 9,

$$y = 0\cdot1627z - 0\cdot0616 \qquad = -0\cdot0044.$$
$$[\beta\beta] = 0\cdot1627\,[\beta\gamma] + 0\cdot5312 = 0\cdot5371.$$
$$[\alpha\beta] = 0\cdot1627\,[\alpha\gamma] - 0\cdot2726 = -0\cdot2784.$$

A	Remarks		Check	T	S	R		z	y	x	
1	—	$[aa]$, $[ab]$, etc.	11·216	0	0	1	−2·588	0·698	4·106	8·000	1
3	—	$[ab]$, $[bb]$, etc.	7·704	0	1	0	−1·442	0·051	3·989	4·106	2
7	—	$[ac]$, $[bc]$, etc.	7·711	1	0	0	1·349	4·613	0·051	0·698	3
2	(1) $\div [aa]$	$\dfrac{[ab]}{[aa]}$, etc.	1·404	—	—	0·125	−0·323	0·087	0·513	…	4
4	(4) $\times [ab]$	$\dfrac{[ab][ab]}{[aa]}$, etc.	5·7648	—	—	0·5132	−1·3262	0·3572	2·1064	…	5
5	(2) − (5)	$[bb1]$, etc.	1·9392	—	1	−0·5132	−0·1158	−0·3062	1·8826	…	6
8	(4) $\times [ac]$	$[ab]\dfrac{[ac]}{[aa]}$, etc.	0·9800	—	—	0·0872	−0·2255	0·0607	0·3581	…	7
9	(3) − (7)	$[bc1]$, $[cc1]$, etc.	6·7310	1	—	−0·0872	1·5735	4·5523	−0·3071	…	8
6	(8) $\div [bb1]$	$\dfrac{[bc1]}{[bb1]}$, etc.	1·0301	—	0·5312	−0·2726	−0·0616	−0·1627	…	…	9
10	(9) $\times [bc1]$	$[bc1]\dfrac{[bc1]}{[bb1]}$, etc.	−0·3163	—	−0·1631	0·0837	0·0189	0·0500	…	…	10
11	(8) − (10)	$[cc2]$, etc.	7·0473	1	0·1631	−0·1609	1·5546	4·5023	…	…	11
12	(11) $\div [cc2]$	$\dfrac{[cl2]}{[cc2]}$, etc.	—	0·2221	0·0362	−0·0357	0·3453	…	…	…	12

From line 4,

$$x + 0\text{·}513y + 0\text{·}087z = -0\text{·}323, \quad \text{and} \quad x = 0\text{·}3517.$$

$$[\alpha\alpha] + 0\text{·}513\,[\alpha\beta] + 0\text{·}087\,[\alpha\gamma] = 0\text{·}125, \quad \text{and} \quad [\alpha\alpha] = 0\text{·}271.$$

Hence the complete solution is:

$$x = 0\text{·}3517 \qquad \text{weight } 3\text{·}68.$$
$$y = -0\text{·}0044 \qquad\qquad \text{,, } \quad 1\text{·}86.$$
$$z = 0\text{·}3453 \qquad\qquad \text{,, } \quad 4\text{·}50.$$

Changing back to the original units

$$x = 351\text{·}7 \qquad p_x = 3\text{·}68.$$
$$y = -4\text{·}4 \qquad p_y = 1\text{·}86.$$
$$z = 345\text{·}3 \qquad p_z = 4\text{·}50.$$

The method here used may be extended to any number of unknowns. If we have to solve a series of normal equations involving a large number of unknowns, where the weights of only a few of the unknowns are required, it is advisable to make these unknowns the last in the order of elimination. If the weights of only two unknowns are required, the solution may be carried out as in table A, page 95, without the addition of extra columns. The coefficient of the last unknown in the final elimination equation will give the weight of that unknown, and the weight of the last unknown but one can be most simply derived by the use of the expression derived on page 107. If it should be required to find the weights of more than two unknowns, it is generally better to introduce an extra column into the table of solution, for each unknown whose weight is required. The weights are then evaluated as part of the general solution of the normal equations. It should be noted that the T column may be omitted from the table of solution for three unknowns, since the weight of the last unknown is the coefficient of that unknown in the final elimination equation.

50. The order of solution shown in the examples worked out in detail in this chapter can be varied to some extent to suit the individual taste of the computer. When the coefficients in the normal equations are given to three or four significant figures (or more), it will generally be found convenient to solve by the Gauss method of substitution, or some variant of that method, particularly if there are more than three unknowns. If there are only three unknowns some computers prefer the determinant method, particularly when the work is carried out by the use of an arithmometer. When the coefficients in the normal equations are small integers the determinant method may be used ; or it may happen that the equations may be more conveniently solved by the methods of elementary algebra, as in Examples 1 and 6 at the end of Chapter V.

The following example illustrates a method of solving the normal equations, and of finding the weights of the unknowns, by simple algebra. A set of four normal equations is given, and the values of the unknowns and their weights are required. The constant terms in the normal equations are written in literal form.

$$3x + 2y + 2z + 2w = [al] \quad\quad\quad\quad (1),$$

$$2x + 3y + 2z + 2w = [bl] \quad\quad\quad\quad (2),$$

$$2x + 2y + 3z + 2w = [cl] \quad\quad\quad\quad (3),$$

$$2x + 2y + 2z + 3w = [dl] \quad\quad\quad\quad (4).$$

Adding all four equations, we find

$$9(x + y + z + w) = [al] + [bl] + [cl] + [dl].$$

Adding together equations (1) and (2),

$$4(x + y + z + w) + x + y = [al] + [bl],$$

or $$x + y = \tfrac{5}{9}[al] + \tfrac{5}{9}[bl] - \tfrac{4}{9}[cl] - \tfrac{4}{9}[dl] \quad\quad\quad (5).$$

Subtracting (2) from (1),

$$x - y = [al] - [bl] \quad\quad\quad\quad\quad (6).$$

From (5) and (6)

$$x = \frac{7[al] - 2[bl] - 2[cl] - 2[dl]}{9},$$

$$y = \frac{-2[al] + 7[bl] - 2[cl] - 2[dl]}{9},$$

with similar forms for z and w.

The values of the unknowns can be deduced by putting into these expressions the arithmetical values of $[al]$, etc. Clearly the weights of all the unknowns are the same, being equal to $\tfrac{7}{9}$.

Example 1. The normal equations solved above (p. 110) were derived from the following set of observational equations :

				Residuals
$1{\cdot}000x$	$-0{\cdot}061y$	$+0{\cdot}907z$	$= -18$	$+20$
$1{\cdot}000$	$-0{\cdot}051$	$+0{\cdot}900$	$= -41$	-1
$1{\cdot}000$	$+0{\cdot}291$	$-0{\cdot}634$	$= -538$	$+34$
$1{\cdot}000$	$+0{\cdot}299$	$-0{\cdot}668$	$= -589$	-5
$1{\cdot}000$	$+0{\cdot}315$	$-0{\cdot}736$	$= -650$	-49
$1{\cdot}000$	$+0{\cdot}999$	$-0{\cdot}817$	$= -139$	-66
$1{\cdot}000$	$+1{\cdot}026$	$+0{\cdot}733$	$= -58$	$+44$
$1{\cdot}000$	$+1{\cdot}288$	$-0{\cdot}621$	$= -549$	$+23$

Verify the residuals given in the last column, and show that the P.E. of an observational equation is $\pm 31{\cdot}4$. Hence show

$$r_x = \pm 16{\cdot}5,$$
$$r_y = \pm 23{\cdot}0,$$
$$r_z = \pm 14{\cdot}8.$$

Example. 2. Find the weight of y in the table above, without using the columns R, S, T. (Vide page 107.)

Would it be possible to find the weight of x in the same way ? If not, what fresh auxiliary products would have to be calculated, in order to yield p_x without using column R ?

51. Alternative Proof of the Rule for finding Weights of the Unknowns.

If r be the P.E. of an observational equation of unit weight, p_x, p_y, p_z, etc., the weights of the unknowns, and r_x, r_y, r_z, etc., the probable errors of the unknowns, then

$$p_x r_x^2 = r^2 = p_y r_y^2 = p_z r_z^2 = \text{etc.}$$

It has already been shown in § 45 that we may write

$$x = \alpha_1 l_1 + \alpha_2 l_2 + \ldots = [\alpha l],$$
$$y = \beta_1 l_1 + \beta_2 l_2 + \ldots = [\beta l],$$
$$z = \gamma_1 l_1 + \gamma_2 l_2 + \ldots = [\gamma l],$$

where the α's, β's, and γ's are constants whose values depend entirely on the values of the coefficients a_1, b_1, c_1, a_2, b_2, c_2, etc., of the observational equations.

Since r is the P.E. of each observed quantity l_1, l_2, etc.,

$$r_x^2 = (\alpha_1^2 + \alpha_2^2 + \ldots)\, r^2 = [\alpha\alpha]\, r^2.$$

$$\therefore\ p_x = \frac{1}{[\alpha\alpha]}.$$

And similarly $p_y = \dfrac{1}{[\beta\beta]}, \qquad p_z = \dfrac{1}{[\gamma\gamma]}.$

We now have to find the values of $[\alpha\alpha]$, $[\beta\beta]$, $[\gamma\gamma]$, etc.

In the subsequent work we shall limit our discussion to the case of three unknowns, but the results will all hold for any number of unknowns.

Substituting $x = [\alpha l]$, $y = [\beta l]$, $z = [\gamma l]$, in the normal equations, we find

$$[aa][\alpha l] + [ab][\beta l] + [ac][\gamma l] - [al] = 0,$$
$$[ab][\alpha l] + [bb][\beta l] + [bc][\gamma l] - [bl] = 0,$$
$$[ac][\alpha l] + [bc][\beta l] + [cc][\gamma l] - [cl] = 0.$$

These relations must be identically true for all values of the l's. It follows that the coefficient of each separate l in each of these equations is zero. Collecting these coefficients and equating them to zero, we obtain three sets of n equations each:

$$\left.\begin{aligned}[aa]\,\alpha_1 + [ab]\,\beta_1 + [ac]\,\gamma_1 - a_1 = 0 \\ [aa]\,\alpha_2 + [ab]\,\beta_2 + [ac]\,\gamma_2 - a_2 = 0 \\ \cdots\cdots\cdots\cdots\cdots\cdots\cdots\cdots\cdots\cdots\cdots \\ \cdots\cdots\cdots\cdots\cdots\cdots\cdots\cdots\cdots\cdots\cdots\end{aligned}\right\}\dots\dots\dots(A_1),$$

$$\left.\begin{aligned}[ab]\,\alpha_1 + [bb]\,\beta_1 + [bc]\,\gamma_1 - b_1 = 0 \\ \cdots\cdots\cdots\cdots\cdots\cdots\cdots\cdots\cdots\cdots\cdots \\ \cdots\cdots\cdots\cdots\cdots\cdots\cdots\cdots\cdots\cdots\cdots\end{aligned}\right\}\dots\dots\dots(A_2),$$

$$\left.\begin{aligned}[ac]\,\alpha_1 + [bc]\,\beta_1 + [cc]\,\gamma_1 - c_1 = 0 \\ \cdots\cdots\cdots\cdots\cdots\cdots\cdots\cdots\cdots\cdots\cdots \\ \cdots\cdots\cdots\cdots\cdots\cdots\cdots\cdots\cdots\cdots\cdots\end{aligned}\right\}\dots\dots\dots(A_3).$$

Multiplying the equations in each set by a_1, a_2, etc., and adding, we obtain the three equations:

$$[aa][a\alpha] + [ab][a\beta] + [ac][a\gamma] - [aa] = 0,$$
$$[ab][a\alpha] + [bb][a\beta] + [bc][a\gamma] - [ab] = 0,$$
$$[ac][a\alpha] + [bc][a\beta] + [cc][a\gamma] - [ac] = 0.$$

This is a set of three equations homogeneous in three unknowns, $[a\alpha] - 1$, $[a\beta]$, and $[a\gamma]$, having no relation between the coefficients; i.e. they are three independent equations. It follows that each variable must vanish.

$$[a\alpha] = 1, \quad [a\beta] = 0, \quad [a\gamma] = 0$$

Similarly it might be shown that

$$\left.\begin{aligned}[a\alpha] &= 1, & [a\beta] &= 0, & [a\gamma] &= 0 \\ [b\alpha] &= 0, & [b\beta] &= 1, & [b\gamma] &= 0 \\ [c\alpha] &= 0, & [c\beta] &= 0, & [c\gamma] &= 1\end{aligned}\right\}\dots\dots\dots(B).$$

Now multiply equations (A_1) by α_1, α_2, etc., and add

$$[aa][\alpha\alpha] + [ab][\alpha\beta] + [ac][\alpha\gamma] = [a\alpha] = 1$$

Similarly from (A_2) and (A_3),

$$[ab][\alpha\alpha] + [bb][\alpha\beta] + [bc][\alpha\gamma] = [b\alpha] = 0 \qquad \ldots\ldots(C_1).$$

$$[ac][\alpha\alpha] + [bc][\alpha\beta] + [cc][\alpha\gamma] = [c\alpha] = 0$$

The equations (C_1) involve three unknowns $[\alpha\alpha]$, $[\alpha\beta]$, $[\alpha\gamma]$. It should be noted that they are the same as the original set of normal equations, with $[\alpha\alpha]$, $[\alpha\beta]$, $[\alpha\gamma]$ substituted for x, y, z, and with $[al] = 1$, $[bl] = [cl] = 0$. Hence we may derive immediately the rule for finding $[\alpha\alpha]$ already given in § 45 above.

$[\alpha\alpha]$ is the value of x derived from the normal equations when

$$[al] = 1, \quad [bl] = 0, \quad [cl] = 0.$$

If equations (A) were multiplied by β_1, β_2, etc., and then by γ_1, γ_2, etc., we should derive two sets of equations $(C_2), (C_3)$, similar to (C_1), from which we should derive the values of $[\beta\beta]$ and $[\gamma\gamma]$.

$[\beta\beta]$ is the value of y derived from the normal equations when

$$[al] = 0, \quad [bl] = 1, \quad [cl] = 0 ;$$

and $[\gamma\gamma]$ is the value of z derived from the normal equations when

$$[al] = 0, \quad [bl] = 0, \quad [cl] = 1.$$

52. Alternative Proof of the Formula

$$r = 0.6745 \sqrt{\frac{[p\,vv]}{n - m}} .$$

For the sake of simplicity in writing we shall consider the case of three unknowns; or $m = 3$.

The observational equations, n in number, are

$$a_1 x + b_1 y + c_1 z - l_1 = v_1 = a_1[al] + b_1[\beta l] + c_1[\gamma l] - l_1,$$

$$a_2 x + b_2 y + c_2 z - l_2 = v_2 = a_2[al] + b_2[\beta l] + c_2[\gamma l] - l_2,$$

$$\text{etc.}$$

$$v_1 = l_1(a_1\alpha_1 + b_1\beta_1 + c_1\gamma_1 - 1) + l_2(a_1\alpha_2 + b_1\beta_2 + c_1\gamma_2) + \text{etc.},$$

$$v_2 = l_1(a_2\alpha_1 + b_2\beta_1 + c_2\gamma_1) + l_2(a_2\alpha_2 + b_2\beta_2 + c_2\gamma_2 - 1) + \text{etc.},$$

$$\text{etc.}$$

If the observations were perfect, the values of l_1, l_2, etc., would be absolutely accurate, and the residuals v_1, v_2, etc., would all be zero. Let dl_1, dl_2, etc., be the errors in l_1, l_2, etc. Then we may write

$$v_1 = (a_1\alpha_1 + b_1\beta_1 + c_1\gamma_1 - 1)\,dl_1 + (a_1\alpha_2 + b_1\beta_2 + c_1\gamma_2)\,dl_2 + \text{etc.}$$

The quantities l_1, l_2, etc., are all determined independently, and so the errors dl_1, dl_2, etc., are all independent, though their mean square errors are all equal. Let this M.S.E. be μ.

Then the mean value of v_1^2 is given by

$$v_1^2 = \mu^2 \left\{ \sum_{r=1}^{r=n} (a_1 a_r + b_1 \beta_r + c_1 \gamma_r)^2 - 2(a_1 \alpha_1 + b_1 \beta_1 + c_1 \gamma_1) + 1 \right\}$$

$$= \mu^2 \{ a_1^2 [\alpha\alpha] + b_1^2 [\beta\beta] + c_1^2 [\gamma\gamma] + 2 a_1 b_1 [\alpha\beta] + 2 a_1 c_1 [\alpha\gamma]$$
$$+ 2 b_1 c_1 [\beta\gamma] - 2(a_1 \alpha_1 + b_1 \beta_1 + c_1 \gamma_1) + 1 \}.$$

We may repeat this process for each residual. Then we obtain the equation

$$[vv] = \mu^2 \{ [aa][\alpha\alpha] + [bb][\beta\beta] + [cc][\gamma\gamma] + 2[ab][\alpha\beta]$$
$$+ 2[ac][\alpha\gamma] + 2[bc][\beta\gamma] - 2[a\alpha + b\beta + c\gamma] + n \}$$

$$= \mu^2 \{ [aa][\alpha\alpha] + [ab][\alpha\beta] + [ac][\alpha\gamma]$$
$$+ [ab][\alpha\beta] + [bb][\beta\beta] + [bc][\beta\gamma]$$
$$+ [ac][\alpha\gamma] + [bc][\beta\gamma] + [cc][\gamma\gamma]$$
$$- 2[a\alpha] - 2[b\beta] - 2[c\gamma] + n \},$$

and referring back to equations (B) and (C) above, we find

$$[vv] = \mu^2 \{ n + 3 - 6 \} = \mu^2 (n - 3).$$

Whence we obtain the usual equation

$$r = 0.6745\mu = 0.6745 \sqrt{\frac{[vv]}{n-3}}.$$

For the case of m unknowns, the equation derived above would be

$$[vv] = \mu^2 \{ m - 2m + n \} = \mu^2 (n - m),$$

and then

$$r = 0.6745 \sqrt{\frac{[vv]}{n-m}} \quad \text{or} \quad 0.6745 \sqrt{\frac{[pvv]}{n-m}}.$$

The working is easily modified to apply to the case where the observational equations have unequal weights.

53. Probable Error of a Function of the Unknowns.

Let the given function be $f(X, Y, Z)$. As in previous work, the function f may be reduced to a linear form,

$$f(X, Y, Z) = f(X_0 Y_0 Z_0) + \frac{\partial f}{\partial X_0} x + \frac{\partial f}{\partial Y_0} y + \frac{\partial f}{\partial Z_0} z,$$

or

$$df = \frac{\partial f}{\partial X_0} x + \frac{\partial f}{\partial Y_0} y + \frac{\partial f}{\partial Z_0} z,$$

where

$$X = X_0 + x, \quad Y = Y_0 + y, \quad Z = Z_0 + z,$$

X_0, Y_0, and Z_0 being approximate values of the unknowns. Since the errors of x, y, z are not independent, we cannot apply to the last equation the reasoning of § 22. With the usual notation,

$$x = [\alpha l], \quad y = [\beta l], \quad z = [\gamma l],$$

and

$$df = \left(\frac{\partial f}{\partial X_0}\right)[\alpha l] + \frac{\partial f}{\partial Y_0}[\beta l] + \frac{\partial f}{\partial Z_0}[\gamma l]$$

$$= \sum_{r=1}^{r=n} \left(\alpha_r \frac{\partial f}{\partial X_0} + \beta_r \frac{\partial f}{\partial Y_0} + \gamma_r \frac{\partial f}{\partial Z_0}\right) l_r.$$

Now the P.E.'s of all the l's are the same, being the P.E. r of an observational equation of unit weight. Hence if r_F be the P.E. of the given function f, we may write

$$r_F^2 = r^2 \sum \left(\alpha_r \frac{\partial f}{\partial X_0} + \beta_r \frac{\partial f}{\partial Y_0} + \gamma_r \frac{\partial f}{\partial Z_0}\right)^2$$

$$= r^2 \left\{ [\alpha\alpha]\left(\frac{\partial f}{\partial X_0}\right)^2 + [\beta\beta]\left(\frac{\partial f}{\partial Y_0}\right)^2 + [\gamma\gamma]\left(\frac{\partial f}{\partial Z_0}\right)^2 \right.$$

$$\left. + 2[\alpha\beta]\frac{\partial f}{\partial X_0}\frac{\partial f}{\partial Y_0} + 2[\beta\gamma]\frac{\partial f}{\partial Y_0}\frac{\partial f}{\partial Z_0} + 2[\alpha\gamma]\frac{\partial f}{\partial X_0}\frac{\partial f}{\partial Z_0} \right\}.$$

It should be noted that in general it is *not* correct to write

$$r_F^2 = \left(\frac{\partial f}{\partial X_0}\right)^2 r_x^2 + \left(\frac{\partial f}{\partial Y_0}\right)^2 r_y^2 + \left(\frac{\partial f}{\partial Z_0}\right)^2 r_z^2.$$

The R.H.S. of the last equation is equivalent to the first three terms on the R.H.S. of the previous equations. In general the other terms do not vanish.

A complete solution of a set of normal equations includes the determination of the weights of the unknowns; i.e. it requires the evaluation of $[\alpha\alpha]$, $[\beta\beta]$, $[\gamma\gamma]$. If the method of the table on page 110 above is followed, the values of $[\alpha\beta]$, $[\beta\gamma]$, $[\alpha\gamma]$ are also derived in the course of the complete solution. It is thus practicable to evaluate r_F from the equation derived here.

54. Normal Place Method in Formation of Observational Equations.

It sometimes happens that a series of observations are such that referred to one of the variable circumstances, say the time, they cluster in groups round certain values of that variable, with well-marked gaps between successive groups. In such a case there is generally no appreciable loss of accuracy in the resulting solution if the mean of each group is taken, and associated with the mean value of the variable (time). Such a mean value is called a *normal place.* The use of the normal place instead of the individual observations reduces the number of observational equations, and produces a considerable saving in time and labour in the formation of the normal equations. For example, if observations of a planet are made at intervals of a fortnight, it is perfectly legitimate to take the mean of one night's observations to form a normal place for working out the elements of the orbit of the planet, provided that the orbit of the planet is not affected by any disturbing cause of short period.

55. Testing the Results of the Least Square Solution for Unusual Errors, and for Systematic or Constant Errors.

When the normal equations have been solved, yielding the values of the unknowns, the next step in the work is to form the residuals by substituting the derived values of the unknowns in the observational equations. The sum of the squares of the residuals will be required in evaluating the probable errors, and even if it should be convenient to evaluate this sum by any other method, the value of $[vv]$ derived directly from the observational equations affords a useful check upon the work of solution. Again it may happen that some of the observations are affected by sources of error not present in the other observations. The effect of such unusual errors would be to yield unusually high values to the corresponding residuals. If some of the residuals should be large in comparison with the probable errors, it may be advisable to reject the observations which yield the high residuals. The problem of the rejection of observations will be considered more fully in Chapter VIII. Meanwhile we note in passing that the evaluation of the residuals is an essential stage in the process of testing the value of the different observations. Further it should be noted that if some of the observations are rejected, it is necessary to repeat the work of solution, as the coefficients in the normal equations are all slightly modified. The values of the unknowns obtained in the first solution may be taken as approximate values, for which the new normal equations will yield corrections.

All the methods of solution hitherto considered are based on the assumption that the observations are affected only by accidental errors, all constant and systematic errors having been removed. The distinction between constant, systematic, and accidental errors has already been discussed in Chapter I above. It was suggested that constant and systematic errors should be eliminated either by changes in the method of solution, or by empirical corrections deduced from special observations designed to determine the exact law of such errors. In practice, however, it is seldom possible to eliminate all the constant and systematic errors, simply because we can never know the nature of all the errors to which an observation is subject. Errors of theory give rise to incorrect coefficients in the observational equations, and these in turn enter into the coefficients of the normal equations, and so affect the values of the unknowns. And when the values of the unknowns are substituted in the normal equations, the errors of theory affect the values of the residuals, and appear in effect as systematic errors.

It is even more important to consider the systematic errors than the accidental errors, since the latter are eliminated by mere repetition, or by the mere increase in the number of observations. The methods of observation should be so arranged as to avoid systematic errors, as far as possible, or to provide corrections for such systematic errors as are not eliminated by the methods of observation. Finally, the residuals may be made to yield information concerning the presence of systematic errors. A number of methods suggest themselves.

1. A comparison of the actual error curve with the theoretical error curve. Thus in § 24, Examples 4 and 5, the curves of actual error show a well-marked deviation from the form of the normal error curve, so suggesting the presence of a systematic error. In Example 4 it was possible to suggest a plausible explanation of this error.

The clustering of points in the diagram representing the observations would also point to the presence of a systematic error.

2. If the residuals show a tendency to have a certain sign when a certain set of conditions exist, and the opposite sign when these conditions are absent, or when other conditions exist, the result may be ascribed to systematic error.

3. When the observations extend over a long period of time, the residuals should be arranged in order of date of observation. If the residuals so arranged follow a systematic law of variation, a systematic error may be expected. For example, an observer knowing nothing of the aberration of light, would find differences between the observed places of stars in different months. But if the observations of one star extended over a number of years, and the errors were arranged in order of date of observation, it would be found that the error for any one star was the same at the same time of the year. The systematic error could then be made an object of prediction, and would cease to be an error.

Again if both the residuals and certain of the conditions of the observations, say the temperature, be arranged in order of date of observation, any correspondence between the variations of the residuals and the conditions considered would indicate a systematic error due to those conditions.

4. A comparison of the results of solution with those of an independent set of observations made under different conditions, or by different methods, may help to determine the presence of constant or systematic errors.

MISCELLANEOUS EXAMPLES.

1. The Hartmann-Cornu formula for the reduction of prismatic spectrograms *.

The usual method of reducing spectrograms, i.e. of measuring the wavelengths of lines in the spectrum by the use of lines of known wave-length, is that due to Hartmann. If n be the measured scale-reading of a line of wave-length λ, n and λ are connected by the formula

$$n - n_0 = \frac{c}{(\lambda - \lambda_0)^a} \quad \dots\dots\dots\dots\dots\dots\dots\dots\dots\dots\dots\dots\dots(1),$$

where n_0, λ_0, and c are constants for each plate, and a is constant for all plates taken with the same instrument. The value of a being known for the particular instrument used, n_0, λ_0, and c can be approximately determined by measuring the positions of three lines of known wave-length. The values of these constants may then be improved by measuring a number of lines, say 12, and giving the constants such values as will afford the best fit to the 12 lines.

* *Vide Monthly Notices, R.A.S.,* Vol. LXXI, p. 663, Stratton.

Corresponding to a line of wave-length λ_r the scale-reading should be

$$n_r = n_0 + \frac{c}{(\lambda_r - \lambda_0)^a} \quad \dots\dots\dots\dots\dots\dots\dots(2).$$

If this equation is not satisfied, let

$$\delta n_r = n_r - \left(n_0 + \frac{c}{(\lambda_r - \lambda_0)^a}\right) \dots\dots\dots\dots\dots\dots(3).$$

We have to find corrections to the constants n_0, λ_0, and c, so that equation (2) shall be satisfied.

The equations to be satisfied for the 12 standard lines are

$$n_r = (n_0 + \partial n_0) + \frac{c + \partial c}{\{\lambda_r - (\lambda_0 + \partial \lambda_0)\}^a} \quad (r = 1, 2, \dots, 12)$$

$$= n_0 + \partial n_0 + \frac{c}{(\lambda_r - \lambda_0)^a} + \frac{\partial c}{(\lambda_r - \lambda_0)^a} + \frac{ac \partial \lambda_0}{(\lambda_r - \lambda_0)^{a+1}}.$$

Making use of equation (3) we may write these equations as

$$\partial n_r = \partial n_0 + \frac{\partial c}{(\lambda_r - \lambda_0)^a} + \frac{ac \partial \lambda_0}{(\lambda_r - \lambda_0)^{a+1}} \quad (r = 1, 2, \dots, 12).$$

These form a set of 12 observational equations which may be solved by the methods of least squares. The normal equations are

$$12 \partial n_0 + \sum_1^{12} \frac{1}{(\lambda_r - \lambda_0)^a} \partial c + \sum_1^{12} \frac{ac}{(\lambda_r - \lambda_0)^{a+1}} \partial \lambda_0 = \sum_1^{12} \partial n_r$$

$$+ \sum_1^{12} \frac{1}{(\lambda_r - \lambda_0)^{2a}} \partial c + \sum_1^{12} \frac{ac}{(\lambda_r - \lambda_0)^{2a+1}} \partial \lambda_0 = \sum_1^{12} \frac{\partial n_r}{(\lambda_r - \lambda_0)^a}$$

$$+ \sum_1^{12} \frac{a^2 c^2}{(\lambda_r - \lambda_0)^{2a+2}} \partial \lambda_0 = \sum_1^{12} \frac{ac \partial n_r}{(\lambda_r - \lambda_0)^{a+1}}.$$

In these equations, all the quantities which occur are known, except ∂n_0, ∂c, $\partial \lambda_0$, for which these equations have to be solved.

In practice, it is found that when the normal equations are solved, the weights of the corrections which they yield are very small, i.e. the corrections are very badly determined. It is customary to assume λ_0 to remain constant, so that $\partial \lambda_0 = 0$. The normal equations then yield corrections ∂n_0 and ∂c which are well-determined.

2.   Position of the sun's axis*.

In an investigation on the position of the sun's axis, Dyson derived his observational equations in the form

$$x \cos \theta_r + y \sin \theta_r + z = l_r,$$

where θ had one of thirteen values, $0°$, $\pm 15°$, $\pm 30°$, $\dots$, etc., $\pm 90°$. Each observed quantity l_r was subject to the same probable error ϵ.

---

* Dyson, *Monthly Notices, R.A.S.*, Vol. LXXII, p. 564.

The normal equations, derived in the usual way, are

$$x\Sigma \cos^2\theta + y\Sigma \sin\theta\cos\theta + z\Sigma\cos\theta = \Sigma l\cos\theta, \quad \text{P.E. } \epsilon\sqrt{\Sigma\cos^2\theta},$$

$$x\Sigma\sin\theta\cos\theta + y\Sigma\sin^2\theta + z\Sigma\sin\theta = \Sigma l\sin\theta, \quad \text{P.E. } \epsilon\sqrt{\Sigma\sin^2\theta},$$

$$x\Sigma\cos\theta + y\Sigma\sin\theta + z13 = \Sigma l, \quad \text{P.E. } \epsilon\sqrt{13},$$

or, evaluating the trigonometric coefficients,

$$6x + 7{\cdot}64z = a, \quad \text{P.E. } \epsilon\sqrt{6},$$

$$7y = \beta, \quad \text{P.E. } \epsilon\sqrt{7},$$

$$7{\cdot}64x + 13z = \gamma, \quad \text{P.E. } \epsilon\sqrt{13}.$$

Thus y is obtained with a P.E. $\dfrac{\epsilon}{\sqrt{7}}$, but it is found from the actual solution that x is determined with a P.E. $2{\cdot}1\epsilon$, so that x is not by any means as clearly determined as y.

The chief value of the investigation leading to the above set of normal equations therefore consists in the determination of the quantity y, which is given by the least squares solution with a P.E. $\dfrac{\epsilon}{\sqrt{7}}$ or $\cdot378\epsilon$. It is of considerable importance to consider how closely this value can be obtained without going through the least squares solution.

Writing down the 13 observational equations in order, and subtracting the 13th from the 1st, the 12th from the 2nd, the 11th from the 3rd, etc., we obtain the equations

$$y = \tfrac{1}{2}(l_1 - l_{13}),$$

$$\cdot966y = \tfrac{1}{2}(l_2 - l_{12}),$$

$$\cdot866y = \tfrac{1}{2}(l_3 - l_{11}),$$

$$\text{etc.}$$

Each of these equations has a P.E. $\dfrac{\epsilon}{\sqrt{2}}$.

Adding the first two of these equations, we obtain $1{\cdot}966y$ with a P.E.

$$\frac{\epsilon}{\sqrt{2}}\sqrt{2}.$$

Adding the first three, we obtain $2{\cdot}832y$ with a P.E. $\dfrac{\epsilon}{\sqrt{2}}\sqrt{3}$; and so on.

In this way y is determined

with P.E. $\cdot707\epsilon$ using one equation at each end of the series of thirteen,

	$\cdot509\epsilon$	,,	two equations	,,	,,	,,	,,
,,	$\cdot432\epsilon$	,,	three	,,	,,	,,	,,
,,	$\cdot409\epsilon$	,,	four	,,	,,	,,	,,
,,	$\cdot392\epsilon$	,,	five	,,	,,	,,	,,
,,	$\cdot396\epsilon$	,,	six	,,	,,	,,	,,
,,	$\cdot378\epsilon$	,,	least squares solution.				

By using five equations at each end of the series we thus obtain a very close approximation to the accuracy of a least squares solution, while four equations at each end only increase the P.E. from $\cdot 38\epsilon$ to $\cdot 41\epsilon$. The use of four equations at each end of the series yields results of a high degree of accuracy, and affords a considerable saving of labour.

The above set of equations illustrates a fact which has wide applications. If in the observational equations two of the unknowns appear with coefficients which are more or less related, e.g. always have the same sign (as in the case of the coefficients of x and z in the above equations), their weights in the solution will be small. If the coefficient of z were always twice that of x we could not determine either separately.

3. Solve the normal equations derived above for x, y, z, and find the weights of x and z. Verify the statement that the P.E. of x is $2\cdot1\epsilon$.

4   Why is it not legitimate to proceed as follows in Example 2 ?

$$13\,(6x + 7\cdot64z) - 7\cdot64\,(7\cdot64x + 13z)$$

has P.E. made up of $13\epsilon\sqrt{6}$ and $7\cdot64\epsilon\sqrt{13}$.

Therefore P.E. of $(78 - 7\cdot64^2)\,x$ is $\sqrt{(13^2.\,6 + 7\cdot64^2.\,13)}$, whence the P.E. of x can be deduced.

CHAPTER VII

THE ADJUSTMENT OF CONDITIONED OBSERVATIONS

56. WHEN the quantities measured, or the unknowns which they involve, are not independent, but are connected *a priori* by certain relations which must be satisfied by the adjusted values, the methods of Chapters V and VI are not directly applicable.

Let there be n directly observed quantities $M_1, M_2, ..., M_n$, of weights $p_1, p_2, ..., p_n$, and let the most probable values of the observed quantities be $L_1, L_2, ..., L_n$.

Then, if v_1, v_2, etc., be the residuals, we have the relations

$$\left. \begin{aligned} L_1 - M_1 &= v_1 \\ L_2 - M_2 &= v_2 \\ \text{etc.} & \end{aligned} \right\} \quad \text{...................(i).}$$

In Chapters V and VI it was supposed that the quantities L_1, L_2, etc., could be accurately represented as functions of certain unknown quantities X, Y, Z, etc., m in number. The same method might be applied to the problem we now have to consider. In general the *a priori* relations mentioned above are expressible as explicit functions of L_1, L_2, etc., and consequently as explicit functions of v_1, v_2, etc.; and it is more convenient to regard

$$v_1, v_2, ..., v_n$$

as n unknown quantities, which are connected by a certain number of conditions, or functional relations. Let these relations, m' in number, be reduced to linear form (cf. page 76), and written

$$\left. \begin{aligned} h_1 v_1 + h_2 v_2 + h_3 v_3 + ... - l_1 &= 0 \\ k_1 v_1 + k_2 v_2 + k_3 v_3 + ... - l_2 &= 0 \\ \text{...................................} & \end{aligned} \right\} \quad \text{...............(ii),}$$

where the coefficients $h_1, h_2, k_1, k_2, l_1, l_2$, etc., are all known quantities. The least square theory requires that $[pvv]$ shall be a minimum subject to the conditions represented by equations (ii). There are two methods of effecting this.

57.   Direct Solution by Substitution.

By the use of equations (ii) it is possible to express m' of the unknown residuals as linear functions of the remaining $n - m'$ residuals. Substituting the values thus obtained in $[pvv]$ we obtain an expression involving only $n - m'$ independent unknowns. Differentiating this expression with respect to each of these independent unknowns in turn, we obtain $n - m'$ linear equations whose solution yields the value of the $n - m'$ independent unknowns. The remaining m' residuals can be evaluated by the use of the expressions first deduced from the equations (ii). The residuals are then all known.

Example 1.   The Adjustment of Coplanar Angles.
The following are the observed values of four coplanar angles. What are the most probable values of the angles?

$$AOB = 80°\ 13'\ 10''\ \text{weight 3,}$$
$$BOC = 83°\ 18'\ 8''\ \quad,, \quad 4,$$
$$COD = 72°\ 6'\ 4''\ \quad,, \quad 2,$$
$$DOA = 124°\ 21'\ 16''\ \quad,, \quad 2.$$
$$\text{Sum} = 359°\ 58'\ 38''.$$

Let x, y, z, w be the corrections, measured in seconds, to be added to the measured values of these angles.

Then 　　　　$359°\ 58'\ 38'' + (x+y+z+w)'' = 360°,$

or 　　　　　　$x+y+z+w = 82$ ....(1).

This conditional equation must be rigorously satisfied by the adjusted values of the angles.
The observational equations are

$$
\left.
\begin{aligned}
x &= 0 \quad \text{weight 3} \\
y &= 0 \quad \quad,, \quad\ 4 \\
z &= 0 \quad \quad,, \quad\ 2 \\
w &= 0 \quad \quad,, \quad\ 2
\end{aligned}
\right\} \quad ....(2).
$$

Substituting for w from equation (1), we replace the last of equations (2) by

$$-x - y - z + 82 = 0 \quad \text{weight 2} \quad ....(3).$$

The normal equations derived from the first three equations in (2), and equation (3), are

$$5x + 2y + 2z = 164,$$
$$2x + 6y + 2z = 164,$$
$$2x + 2y + 4z = 164.$$

Subtracting the second equation from the first, and the third from the second, we find

$$3x - 4y = 0, \qquad 4y - 2z = 0.$$

Whence we find

$$x = \tfrac{328}{19}, \quad y = \tfrac{246}{19}, \quad z = \tfrac{492}{19}, \quad w = \tfrac{492}{19},$$

or
$$x = 17'', \quad y = 13'', \quad z = 26'', \quad w = 26''.$$

The adjusted values of the angles are therefore

$$AOB = 80°\ 13'\ 27'',$$
$$BOC = 83°\ 18'\ 21'',$$
$$COD = 72°\ 6'\ 30'',$$
$$DOA = 124°\ 21'\ 42''.$$

Example 2. The three angles of a spherical triangle are all equally well observed. To find the most probable values of the angles.

Let A, B, C be the three angles, and ϵ the spherical excess of the triangle.

Then
$$A + B + C = 180° + \epsilon.$$

Let M_1, M_2, M_3 be the observed values of the angles, and let

$$A = M_1 + v_1,$$
$$B = M_2 + v_2,$$
$$C = M_3 + v_3.$$

Then the problem is to find the most probable values of the residuals v_1, v_2, v_3,

$$v_1 + v_2 + v_3 = 180° + \epsilon - M_1 - M_2 - M_3 = a, \text{ say.}$$

The least square theory requires that

$$v_1{}^2 + v_2{}^2 + v_3{}^2$$

shall be a minimum subject to the condition

$$v_1 + v_2 + v_3 = a.$$

Substituting for v_3, we find that

$$v_1{}^2 + v_2{}^2 + (a - v_1 - v_2)^2$$

must be a minimum.

Differentiating with respect to v_1 and v_2 in turn, we obtain the equations

$$2v_1 + v_2 = a,$$
$$v_1 + 2v_2 = a.$$

Whence we deduce

$$v_1 = v_2 = \frac{a}{3} = v_3.$$

58. Method of Undetermined Multipliers or Correlates.
The method of direct substitution is only practicable when the relations expressed by equations (ii) are few in number and simple in form. In other cases the minimum value of $[pvv]$ can best be found by the method of "Undetermined Multipliers."

Equations (ii) are multiplied by $-2A_1$, $-2A_2$, etc., and the products are added to $[pvv]$, yielding the expression

$$[pvv] - 2A_1(h_1v_1 + h_2v_2 + \ldots - l_1) - 2A_2(k_1v_1 + k_2v_2 + \ldots - l_2) \text{ etc. (iii)}.$$

We then proceed to find the values of v which will make this expression a minimum. Differentiating with respect to v_1, v_2, etc. in turn, we obtain a system of n equations:

$$\left.\begin{aligned} p_1v_1 &= A_1h_1 + A_2k_1 + \ldots \\ p_2v_2 &= A_1h_2 + A_2k_2 + \ldots \\ & \end{aligned}\right\} \quad \ldots\ldots\ldots\ldots\ldots\ldots(\text{iv}).$$

Substituting these values of v_1, v_2, etc. in the equations (ii), we obtain a system of m' equations:

$$\left.\begin{aligned} A_1\left[\frac{hh}{p}\right] + A_2\left[\frac{hk}{p}\right] + \ldots &= l_1 \\ A_1\left[\frac{hk}{p}\right] + A_2\left[\frac{kk}{p}\right] + \ldots &= l_2 \end{aligned}\right\} \quad \ldots\ldots\ldots\ldots(\text{v}),$$

where

$$\left[\frac{hh}{p}\right] = \frac{h_1^2}{p_1} + \frac{h_2^2}{p_2} + \ldots \text{ etc.},$$

$$\left[\frac{hk}{p}\right] = \frac{h_1k_1}{p_1} + \frac{h_2k_2}{p_2} + \ldots \text{ etc.}$$

These m' equations involve the m' correlates A_1, A_2, etc. as unknowns, and the solution yields the values of the correlates. Substituting the values so obtained in equations (iv), we obtain the most probable values of the residuals v_1, v_2, etc. Equations (v) are called the normal equations for the correlates A_1, A_2, etc.

Example. Example 1 above might be worked out by this method. We should have to make

$$3x^2 + 4y^2 + 2z^2 + 2w^2 - 2A(x + y + z + w - 82)$$

a minimum.

Differentiating with respect to x, y, z, w, in turn, we find

$$3x = A = 4y = 2z = 2w.$$

Since $x + y + z + w = 82,$

we immediately deduce the same values of x, y, z, w as were given above.

59.   The Precision of the Unknowns.

It was shown on page 100 that in a system of n observations involving m independent unknowns, the P.E. of an observational equation of unit weight is

$$r = 0\text{·}6745 \sqrt{\frac{[pvv]}{n-m}}.$$

In the case considered in this chapter there are n unknowns, connected by m' relations.   But we can use the m' relations (ii) to eliminate m' of the unknown quantities, so that we shall have n observed values, involving $n - m'$ independent unknowns.   We may then use the formula given above for evaluating r.

But $\qquad\qquad m = n - m'$ or $n - m = m'$,

$$\therefore\ r = 0\text{·}6745 \sqrt{\frac{[pvv]}{m'}}.$$

The P.E. of a residual v_s of weight p_s is

$$r_s = 0\text{·}6745 \sqrt{\frac{[pvv]}{m'p_s}}.$$

For a full development of the subject of this chapter, and in application to survey work in particular, the reader is referred to

Wright and Hayford, *Adjustment of Observations*, Chapters v, vi, and vii.

F. R. Helmert, *Ausgleichungsrechnung*, Chapter iv.

Oscar S. Adams, *Application of Theory of Least Squares to the Adjustment of Triangulation*.   U.S. Coast and Geodetic Survey, Special Publication No. 28.

Jordan, *Handbuch der Vermessungskunde*, Bd. i.

CHAPTER VIII

THE REJECTION OF OBSERVATIONS

60. An observer making a series of observations of any kind has power to reject any observation if he is certain that it is vitiated by some unusual sources of error which do not affect the other observations in the series. To put this in other words, the observer is to a certain extent free to choose the time for making his observations so that the external conditions shall vary as little as possible during the series. He is supreme in his own department, having the power to retain or reject observations according to his judgment of the extent to which the external causes of error react upon his measurements.

But when the observational material is put into the hands of the computer there arises a new question. Shall the computer be allowed to reject any observation whose residual is much larger than those of the remaining observations? He should clearly be allowed to reject an observation when he is convinced that it is affected by an error from some unusual source, which does not affect the other observations in the series; or if the observation is clearly spoiled by some definite blunder, such as the mis-reading of a scale by five divisions. It is sometimes possible to correct blunders of this type, and to retain the observation, but the greatest caution is necessary in making such corrections. The real difficulty lies in deciding whether a large residual is due to some unusual source of error, or is due to the chance occurrence of a large number of small accidental errors with the same sign. If the latter alternative be true, the large residual is in accordance with the law of error, and its rejection may decrease rather than increase the accuracy of the final result.

A number of criteria, based on more or less rigid analysis, have
been put forward by various writers. The best known of these is
Peirce's Criterion. The underlying principle is that the doubtful
observations should be rejected when the probability of the system
of errors obtained by retaining them is less than the product of
the probability of the system of errors obtained by rejecting them,
multiplied by the probability of making so many, and no more,
abnormal observations. As this criterion is not in general use,
and is rather tedious in its application, we shall not enter into the
full proof. The reader who desires to know more of it is referred
to Chauvenet's *Theoretical and Practical Astronomy*, Vol. II,
Appendix, § 58, where Peirce's proof is reproduced almost word
for word; and to the proof and tables by Gould in *Astronomical
Journal*, Vol. IV, and *U.S. Coast and Geodetic Survey Report*, 1854,
pp. 131, 132*.

Chauvenet (Vol. II) gives a criterion "for the rejection of one
doubtful observation." The probability that an error is less than
t is, as we have already seen,

$$\frac{2}{\sqrt{\pi}}\int_0^{\frac{t}{h}} e^{-t^2}\, dt = \Theta(t) = \frac{2}{\sqrt{\pi}}\int_0^{\frac{\rho t}{r}} e^{-t^2}\, dt.$$

Of a series of m observations, the number whose errors may be
expected to be less than t will be $m\,\Theta(t)$, and the number which
will exceed t will be $m\,(1 - \Theta(t))$. If this last quantity be less
than $\frac{1}{2}$, then it follows that an error greater than t has a greater
probability against it than for it, and may therefore be rejected.
The limiting error t which may be rejected is therefore given by

$$m\,(1 - \Theta(t)) = \tfrac{1}{2} \quad \text{or} \quad \Theta(t) = \frac{2m - 1}{2m}.$$

The function $\Theta(t)$ has been tabulated on page 19.

The application of the criterion by means of the formula given
above is extremely simple. Suppose we have a series of 100
observations. Then

$$\frac{2m - 1}{2m} = {\cdot}995.$$

Referring to page 19 we find this value for $\Theta(t)$ when $\dfrac{t}{r} = 4 {\cdot} 2$. The

---

* See also, "Note on Peirce's Criterion," S. A. Saunder, *Monthly Notices, R.A.S.*,
Vol. LXIII, p. 432.

limiting value of t is 4·2 times the probable error. Hence if one of a series of 100 observations has a residual greater than 4·2 times the probable error of the series, Chauvenet's criterion will reject that observation.

Chauvenet's criterion only considers the rejection of one observation, but when one has been rejected, the rule may be applied to consider the rejection of another, and so on. The criterion is simple and easy to apply, but it is probably too sweeping. It does not allow sufficiently for the possible presence of a large number of small accidental errors of the same sign.

Other criteria have been suggested, but few of them are really useful in practice. The whole question of the possibility of rejecting observations on the ground of theoretical discussion based on residuals only, has given rise to a considerable amount of controversy. Bessel opposed the rejection of any observation unless the observer was satisfied that the external conditions produced some unusual source of error not present in the other observations of the series. Peirce's criterion was at an early date subjected to very severe criticism. Airy* claimed that it was defective in foundation, and illusive in its results. He maintained that, so long as the observer was satisfied that the same sources of error were at work, though in varying degrees, throughout a series of observations, the computer should have no right to reject any observation by a discussion based solely on antecedent probability. An observation should be rejected only when a thorough examination showed that the causes of error normally at work were not sufficient to produce the error in the doubtful observation. Airy also cited a case where the rejection of the observations having large residuals led to poor results. In the preceding century, at a time when the figure of the earth was not well determined, azimuth observations were made at Beachy Head and Dunnose, in connection with the survey of England, and the results obtained were poor. Later on, when the full record of the observations fell into the hands of General Colby, it was found that only the observations which were in closest agreement had been used in the reduction. When the calculations were repeated, using all the observations, results of a high degree of accuracy were obtained.

* *Astronomical Journal*, Vol. IV, p. 137.

Though many of the arguments of Airy and others against the use of mathematical criteria such as Peirce's have been shown to be based on faulty premises, the fact remains that none of these criteria have ever come into general use.

If the distribution of errors be in strict accordance with the law of error, only a few large residuals will occur. Referring to the table on page 19, we see that the probability of an error greater than $5r$, where r is the probable error of a single observation, is ·001, and the probability of an error greater than $3·5r$ is ·018. Thus only one observation in 1000 should have an error as great as $5r$, while one error in 55 should have an error as great as $3·5r$. These numbers form the theoretical basis of the following rule, which is advocated for general use by Wright and Hayford*.

"Reject each observation for which the residual exceeds five times the probable error of a single observation. Examine carefully each observation for which the residual exceeds $3·5$ times the probable error, and reject it if any of the accompanying conditions are such as to produce lack of confidence."

This criterion has the merits of simplicity, ease of application, and a fairly sound theoretical basis. For a moderate number of observations it cannot be said to be too sweeping.

* *Adjustment of Observations.*

CHAPTER IX

61. WE have already dealt briefly with one or two cases in which the frequency distribution showed a well-marked deviation from the normal form. In economic and biological statistics, in particular, the frequency distributions are liable to show considerable dis-symmetry. It is therefore necessary to consider the methods of representing such frequency distributions by some substitute for the normal error curve of Gauss. The purpose which has to be kept in view is that of replacing the series which represents the observations by a simple formula involving a few constants only.

Before proceeding to the development of possible formulae which shall represent various types of frequency distribution, we must define briefly a number of statistical terms. Let the data be arranged in the form of a frequency distribution, and let f_x be the frequency of the *characteristic* x. The *characteristic* is the scale in terms of which the observations are made. In the diagrams of the earlier chapters of this book, it is the variable represented along the horizontal axis (the x-axis).

The *median* is the value of the characteristic which has as many observations on one side of it as on the other.

The *mode* is the value of the characteristic corresponding to the maximum ordinate of the frequency curve. The position of this ordinate cannot be accurately determined until the form of the frequency curve is known, since it is not necessarily the ordinate corresponding to the biggest number of actual observations. (Compare figure 7, page 46.)

The *mean* is the average value of the characteristic, and is the arithmetic mean of all the observed values. If f_x be the frequency of a value x of the characteristic, or the frequency of a group centred round x, the mean is

$$\frac{\Sigma f_x x}{\Sigma f_x},$$

or, for the curve,
$$\frac{\int f_x\, x\, dx}{\int f_x\, dx}.$$

The ordinate through the mean is often spoken of as the *centroid vertical*, since it passes through the centre of gravity of the distribution.

The *standard deviation*, or S.D., represented by the symbol σ, measures the closeness with which the measurements are clustered about the mean. Measuring x from the mean, σ is given by

$$\sigma^2 = \frac{\Sigma x^2 f_x}{\Sigma f_x} \quad \text{or} \quad \sigma^2 = \frac{\int x^2 f_x dx}{\int f_x dx}.$$

The first or second form is to be used according as the calculation is made from the observations or from the curve. Using the notation of Chapter III, we should have

$$\sigma^2 = \frac{[vv]}{n} \quad \text{where } n = \Sigma f_x.$$

The mean square error is given by

$$\mu^2 = \frac{[vv]}{n-1}.$$

Thus $\sigma = \mu \sqrt{\dfrac{n-1}{n}}$, and when n is large, σ and μ may be regarded as identical. For the present we shall call this quantity the S.D. (σ), rather than the M.S.E. (μ), since it is customary to do so in all works on general statistics. It should be noted that σ does not depend upon the actual frequencies, but only on their distribution. It measures the scatter of the observations about the mean.

When a curve is symmetrical, the mean, mode, and median coincide. An unsymmetrical curve is often described as a *skew* curve. The *skewness* of a curve or of a frequency distribution is measured by the distance between the mean and the mode. It is convenient to define the skewness by the equation

$$\text{Skewness} = \frac{\text{Mean} - \text{Mode}}{\text{S. D.}},$$

so that a curve with *positive skewness* has the mode to the left of the mean in the usual diagram, or in other words, when the rise to the maximum is more rapid than the fall from the maximum. In a skew curve the mode is the value of the characteristic which has the greatest frequency.

The nth *moment* of a frequency distribution about any ordinate is obtained by multiplying each frequency by the nth power of its distance from the ordinate in question, and adding together all the products. With our present notation, the nth moment about the ordinate $x = a$ is

$$\Sigma f_x (x - a)^n.$$

It is customary to employ the symbol μ_n to represent the nth moment about the mean, and the accented symbol μ_n' to denote the nth moment about any other ordinate. It is generally more convenient to evaluate moments about some other ordinate than the mean, e.g. an ordinate corresponding to an integral value of the characteristic, and then to deduce from these moments the corresponding moments about the mean. Let the distance between the assumed ordinate and the mean be a, and let the distance from any other ordinate to these two ordinates be x_r' and x_r respectively. Then $x_r = x_r' - a$.

$$\mu_n = \Sigma f_r (x_r' - a)^n$$

$$= \Sigma f_r x_r'^n - na \, \Sigma f_r x_r'^{n-1} + \text{etc.}$$

$$= \mu_n' - na\mu'_{n-1} + \frac{n.(n-1)}{1.2} a^2\mu'_{n-2} - \text{etc.} \dots\dots\dots(1).$$

Or, we may write

$$\mu_n' = \Sigma f_n (x_r')^n = \Sigma f_r (x_r + a)^n$$

$$= \mu_n + na\mu_{n-1} + \frac{n.n-1}{1.2} a^2\mu_{n-2} + \dots.$$

Whence it follows that

$$\mu_n = \mu_n' - na\mu_{n-1} - \frac{n.n-1}{1.2} a^2 \mu_{n-2} - \text{etc.}\dots\dots\dots(2).$$

Either of equations (1) and (2) may be used to evaluate the moments in turn, the two equations of necessity leading to identical results.

62. Most of the methods of fitting curves other than the normal error curve, to series of observations, are based on the use of moments. If the functional form contains n constants, these are deduced by making the moments deduced from the observations agree with the moments deduced from the curve, up to the nth moment. This process yields n equations which suffice to determine the n constants. The following simple example may help to illustrate the utility of the method.

It is required to fit a curve of the form

$$y = a + bx + cx^2$$

to the frequency distribution

$x = 1$	$y = 10$
2	12
3	18
4	36

If only the first three terms corresponding to $x = 1, 2$, and 3, were given, then by substituting the values of x and y in the assumed equation we could solve the resulting three equations, and so obtain a, b, and c. The results so obtained are

$$a = 12, \qquad b = -4, \qquad c = 2.$$

Thus the curve which accurately fits the first three terms is

$$y = 12 - 4x + 2x^2.$$

But when we substitute $x = 4$ in this equation we find $y = 28$. Thus the curve does not fit the fourth term $x = 4$, $y = 36$. If any three out of the four points given above were selected, a curve could be made to pass through these three points, but it would not pass through the fourth point. It is therefore necessary to adopt some method of calculating the constants a, b, c, so that all four points will be taken into account. The result will be to yield a curve which will pass very near to all four points, but which need not of necessity pass through any one of them. This could easily be done by the method of least squares. Another simple method is to adopt values of the constants a, b, c, such that the first three moments of the actual distribution about the origin shall be equal to the first three moments yielded by the curve.

Taking the zero moment, and the first and second moments, we obtain three equations:

$$(a+b+c)+ \quad (a+2b+2^2c)+ \quad (a+3b+3^2c)+ \quad (a+4b+4^2c)$$
$$= 10+12+18+36,$$

$$1\ (a+b+c)+2\ (a+2b+2^2c)+3\ (a+3b+3^2c)+4\ (a+4b+4^2c)$$
$$= 10.1+12.2+18.3+36.4,$$

$$1^2\,(a+b+c)+2^2\,(a+2b+2^2c)+3^2\,(a+3b+3^2c)+4^2\,(a+4b+4^2c)$$
$$= 10.\ 1^2+12.\ 2^2+18.\ 3^2+36.\ 4^2,$$

or,

$$4a+\ 10b+\ 30c=\ 76,$$
$$10a+\ 30b+100c=232,$$
$$30a+100b+354c=796.$$

The solution of these equations yields

$$a=18, \qquad b=-11\cdot6, \qquad c=4.$$

The resulting curve is

$$y=18-11\cdot6x+4x^2.$$

x	Calculated y	Observed y
1	10·4	10
2	10·2	12
3	19·2	18
4	35·6	36

The table shows that the adopted curve does not fit any of the observations accurately, but yields values of y which are sometimes greater, sometimes less, than the observed values.

This simple example may help to illustrate the general nature of statistical problems, where it is necessary to replace the irregularities of observations by a smooth curve.

63. Pearson's Curves.

The general theory of curve-fitting has been worked out in great detail by Prof. Karl Pearson*. Starting from certain properties which may be considered essential in good observations, Pearson has derived a series of formulae for possible curves of presumptive errors.

(1) The expression must replace the rough material of observation by a smooth continuous curve; i.e. it must graduate the observations.

(2) The expression must not involve too many constants, and those present must be calculable from the material of observation.

(3) There must be a systematic method of approaching frequency distributions.

(4) If the material is homogeneous, the ordinate of the curve will start from zero, increase to a maximum, and fall again, possibly at a different rate, to zero.

(5) The frequency curve will generally have contact with the axis of x at the ends of the range.

Of these, (1), (2) and (3) call for no remark, while (4) and (5) are properties which are generally associated with the frequency distributions obtained in actual practice. The conditions are in general satisfied by the curve whose equation is

$$\frac{1}{y}\frac{dy}{dx} = \frac{x+a}{f(x)}.$$

For, $\frac{dy}{dx} = 0$ when $y = 0$, so that the curve has contact at the axis, and $\frac{dy}{dx} = 0$ when $x = -a$, corresponding to a maximum of the curve.

Expanding by Maclaurin's Theorem, we may write

$$f(x) = b_0 + b_1 x + b_2 x^2 + \ldots \text{ etc.}$$

Pearson has considered in detail the case where $f(x)$ is limited to the first three terms in the expansion. Four constants are then involved, requiring the evaluation of four moments. In general there is little advantage in considering formulae which require

---

* *Phil. Trans. R.S.* 186 A, p. 343; 216 A, p. 429; *Biometrika*, v, p. 172.

the evaluation of higher moments, since the higher moments are very sensitive to errors in the frequency distribution. We shall therefore start from the differential equation

$$\frac{1}{y}\frac{dy}{dx} = \frac{x+a}{b_0 + b_1 x + b_2 x^2} \quad\ldots\ldots\ldots\ldots\ldots(3).$$

Pearson's method consists essentially in making the moments calculated from the curve equal to the moments derived from the observations. The differential equation may be written

$$(b_0 + b_1 x + b_2 x^2)\frac{dy}{dx} = y\,(x+a).$$

Multiplying each side by x^n, and integrating, we find

$$\int x^n (b_0 + b_1 x + b_2 x^2)\frac{dy}{dx}\,dx = \int y\,(x+a)\,x^n\,dx.$$

Integrating the L. H. S. by parts, treating $\frac{dy}{dx}$ as one part, we find

$$x^n (b_0 + b_1 x + b_2 x^2)\,y - \int \{nb_0 x^{n-1} + (n+1)\,b_1 x^n + (n+2)\,b_2 x^{n+1}\}\,y\,dx$$

$$= \int y\,x^{n+1}\,dx + a\int y\,x^n\,dx.$$

But since y becomes zero at each end of the range, we find, in terms of our previous notation,

$$- nb_0\mu'_{n-1} - (n+1)\,b_1\,\mu_n' - (n+2)\,b_2\mu'_{n+1}$$

$$= \mu'_{n+1} + a\mu_n',$$

or $a\mu_n' + nb_0\,\mu'_{n-1} + (n+1)\,b_1\,\mu_n' + (n+2)\,b_2\,\mu'_{n+1} = -\,\mu'_{n+1}.$

Putting $n = 0,\ 1,\ 2,\ 3$, in turn, in this equation, we might obtain four equations to determine the four constants $a,\ b_0,\ b_1,\ b_2$. These equations were derived independently of any assumption as to the position of the origin. If the origin be at the mean, the accents may be omitted, and μ_1 may be equated to zero. The four equations then become

$$
\begin{aligned}
n &= 0 & a + b_1 &= 0 \\
n &= 1 & b_0 + 3b_2\mu_2 &= -\mu_2 \\
n &= 2 & a\mu_2 + 3b_1\,\mu_2 + 4b_2\mu_3 &= -\mu_3 \\
n &= 3 & a\mu_3 + 3b_0\mu_2 + 4b_1\mu_3 + 5b_2\mu_4 &= -\mu_4
\end{aligned}
\quad\Bigg\}\ \ldots\ldots(4).
$$

Sheppard's Corrections. The μ's in these equations represent moments calculated from the curve. In practice, however, the moments are calculated from grouped frequencies, where all values of the characteristic within a certain interval are regarded as equal to the value at the centre of that interval. The moments so calculated are not precisely the same as the moments calculated from the curve. If ν_1, ν_2, etc. be the moments calculated from grouped frequencies, Sheppard * has shown that when there is high contact at the axis, certain corrections must be applied to the moments. The relations between the various moments are shown by the following equations:

$$\left.\begin{aligned} \mu_1 &= \nu_1 \\ \mu_2 &= \nu_2 - \tfrac{1}{12} \\ \mu_3 &= \nu_3 \\ \mu_4 &= \nu_4 - \tfrac{1}{2}\nu_2 + \tfrac{7}{240} = \nu_4 - \tfrac{1}{2}\mu_2 - \tfrac{1}{80} \end{aligned}\right\} \dots\dots\dots(5).$$

Using accented ν's to denote the moments of the grouped observations about other ordinates than the centroid vertical, the relations (1) and (2) above hold for ν_n, ν_n', etc. There are thus three stages in the evaluation of μ_1, μ_2, etc.

(1) Evaluate ν_n' for $n = 1$, 2, 3, etc. about any convenient ordinate.

(2) Transform to the centroid vertical by using equations (1) or (2), so obtaining ν_2, ν_3, etc. $\nu_1 = 0$.

(3) From these values deduce μ_2, μ_3, μ_4, etc. by the use of equations (5) above. It should be remembered that, referred to the centroid vertical, $\mu_1 = \nu_1 = 0$.

The solution of equations (4) for a, b_0, b_1, b_2 is quite straightforward. The resulting form of equation (3) is

$$-\frac{1}{y}\frac{dy}{dx} = \frac{x + \dfrac{\mu_3(\mu_4 + 3\mu_2{}^2)}{10\mu_2\mu_4 - 18\mu_2{}^3 - 12\mu_3{}^2}}{\dfrac{\mu_2(4\mu_2\mu_4 - 3\mu_3{}^2) + \mu_3(\mu_4 + 3\mu_2{}^2)x + (2\mu_2\mu_4 - 3\mu_3{}^2 - 6\mu_2{}^3)x^2}{10\mu_2\mu_4 - 18\mu_2{}^3 - 12\mu_3{}^2}}.$$

---

* W. F. Sheppard, *Proc. L.M.S.* xxix, p. 353; see also, Karl Pearson, *Biometrika*, iii, p. 308.

If in this last form we substitute

$$\beta_1 = \frac{\mu_3{}^2}{\mu_2{}^3}, \qquad \beta_2 = \frac{\mu_4}{\mu_2{}^2}, \qquad \sigma = \sqrt{\mu_2},$$

we obtain the equation in the form

$$-\frac{1}{y}\frac{dy}{dx} = \frac{\sqrt{\beta_1}\,(\beta_2+3) + (10\beta_2 - 12\beta_1 - 18)\dfrac{x}{\sigma}}{\sigma\left\{(4\beta_2 - 3\beta_1) + \sqrt{\beta_1}\,(\beta_2+3)\dfrac{x}{\sigma} + (2\beta_2 - 3\beta_1 - 6)\dfrac{x^2}{\sigma^2}\right\}} \quad \dots(6).$$

This equation is referred to the mean as origin. The mode is obtained by making the numerator in the R.H.S. zero. It follows that the skewness of the curve is

$$\frac{\sqrt{\beta_1}}{2} \cdot \frac{\beta_2+3}{5\beta_2 - 6\beta_1 - 9}.$$

The curve is symmetrical when $\beta_1 = 0$.

The form of the curve is fixed by the nature of the roots of the equation

$$b_0 + b_1 x + b_2 x^2 = 0,$$

i.e. by the value of $b_1{}^2 - 4b_0 b_2$; or by $\dfrac{b_1{}^2}{4b_0 b_2}$.

Substituting the values of b_0, b_1, b_2 derived above, we find

$$\kappa = \frac{b_1{}^2}{4b_0 b_2} = \frac{\beta_1(\beta_2+3)^2}{4(2\beta_2 - 3\beta_1 - 6)(4\beta_2 - 3\beta_1)} \quad \dots\dots(7).$$

The value of this function of the moments fixes the nature of the curve of frequencies. Hence it is known as the *criterion*.

64. Integration of the Differential Equation.

Pearson treats the distribution of the observations as given by N, $\bar{x}$, σ, β_1 and β_2. The form of the frequency curve which satisfies equation (6) is determined by the values of β_1 and β_2, these quantities being positive and non-dimensional, and independent of the units in which x is measured. For different possible distributions β_1 and β_2 can occur anywhere within the range 0 to ∞, and each point in the positive quadrant of the $\beta_1\beta_2$ plane appears to be capable of representing a distribution.

Thus there would appear to be a doubly infinite number of possible curves. Pearson has however shown* that from the algebraic definition of β_1 and β_2 it follows that $\beta_2 - \beta_1 - 1$ is always positive, and the points in the $\beta_1 \beta_2$ plane which are significant are in the region between the line

$$\beta_2 - \beta_1 - 1 = 0$$

and the axis of β_2.

It should also be noted in equation (7) that in consequence of the restriction $\beta_2 > \beta_1 + 1$, the factor $4\beta_2 - 3\beta_1$ is always positive, so that the sign of κ will depend only on the sign of $2\beta_2 - 3\beta_1 - 6$.

Returning to the general differential equation

$$\frac{1}{y}\frac{dy}{dx} = \frac{x + a}{b_0 + b_1 x + b_2 x^2},$$

we find that there are three main types of solutions, according as the roots of the equation

$$b_0 + b_1 x + b_2 x^2 = 0$$

are (1) real and of opposite sign,

(2) imaginary,

or (3) real and of like sign.

These correspond to

(1) $\kappa < 0$,

(2) $0 < \kappa < 1$,

and (3) $\kappa > 1$.

In the positive quadrant of the $\beta_1 \beta_2$ plane each of these three possibilities is represented by an area, separated by the lines corresponding to $\kappa = 0$ and $\kappa = 1$. There will be a number of transitional types and special cases, corresponding to lines or points in the $\beta_1 \beta_2$ quadrant, but the three cases (1), (2) and (3) cover the whole of the possible area bounded by the lines

$$\beta_2 - \beta_1 - 1 = 0 \quad \text{and} \quad \beta_1 = 0,$$

apart from the lines or points to which transitional or special types are appropiate. We shall consider these in turn.

* *Phil. Trans. Roy. Soc.* 216 A, pp. 429–457.

(1) Roots real and of opposite sign. Pearson's Type I.

The differential equation may be written

$$\frac{b_2}{y}\frac{dy}{dx} = \frac{x+a}{(x+c_1)(x-c_2)} = \frac{c_1-a}{c_1+c_2}\frac{1}{x+c_1} + \frac{c_2+a}{c_1+c_2}\frac{1}{x-c_2}.$$

Integration of this equation gives

$$b_2 \log y = \frac{c_1-a}{c_1+c_2}\log(x+c_1) + \frac{c_2+a}{c_1+c_2}\log(x-c_2).$$

When the origin is removed to $x=-a$, and the following substitutions made,

$$a_1 = c_1-a, \qquad a_2 = c_2+a, \qquad p = \frac{1}{b_2(c_1+c_2)},$$

the equation becomes

$$y = y_0\left(1+\frac{x}{a_1}\right)^{pa_1}\left(1-\frac{x}{a_2}\right)^{pa_2}.$$

The curve is limited in both directions, and is skew. It is usually bell-shaped, but may be U-shaped or J-shaped.

(2) Roots imaginary. Pearson's Type IV.

The equation may be simplified slightly by transferring the origin to $x=-\frac{b_1}{2b_2}$, and substituting $c=a-\frac{b_1}{2b_2}$ and $d^2=\frac{b_0}{b_2}-\frac{b_1{}^2}{4b_0b_2}$.

It then becomes

$$\frac{1}{y}\frac{dy}{dx} = \frac{x+c}{b_2(x^2+d^2)} = \frac{1}{b_2}\frac{x}{x^2+d^2} + \frac{c}{b_2}\frac{1}{x^2+d^2},$$

which yields on integration

$$\log y = \frac{1}{2b_2}\log(x^2+d^2) + \frac{c}{b_2 d}\tan^{-1}\frac{x}{d} + \text{constant},$$

or

$$y = y_0\left(1+\frac{x^2}{d^2}\right)^{-m}e^{-p\tan^{-1}\frac{x}{d}}.$$

The curve is skew and unlimited in both directions. It is bell-shaped.

(3) The roots are real and of like sign. Pearson's Type VI.

Let the roots be c_1 and c_2. The result follows as in (1) above,

$$b_2 \log y = \frac{c_1+a}{c_1-c_2}\log(x-c_1) - \frac{c_2+a}{c_1-c_2}\log(x-c_2) + \text{const}.$$

Changing the origin to $x=-c_1$, we may write

$$y = y_0\, x^{m_1}(x-c_1)^{-m_2}.$$

The curve has unlimited range in one direction. It is usually bell-shaped, but may be J-shaped.

Transitional or Special Cases.

(4) Transition from (2) to (3). $\kappa = 1$. Pearson's Type V. The equation has real and equal roots, and may be written

$$\frac{1}{y}\frac{dy}{dx} = \frac{x+a}{b_2\left(x+\frac{b_1}{2b_2}\right)^2} = \frac{1}{b_2}\frac{1}{x+\frac{b_1}{2b_2}} + \frac{1}{b_2}\frac{a-\frac{b_1}{b_2}}{\left(x+\frac{b_1}{2b_2}\right)^2}.$$

Integrating this and removing the origin to $x = -\frac{b_1}{2b_2}$ and writing

$$p = \frac{1}{b_2}\left(a - \frac{b_1}{2b_2}\right), \qquad q = \frac{1}{b_2},$$

we find
$$y = y_0 x^q e^{-\frac{p}{x}}.$$

This is a skew curve with limited range in one direction.

(5) Transition from (3) to (1). $\kappa = \infty$. Pearson's Type III. The curve corresponds to $2\beta_2 - 3\beta_1 - 6 = 0$, or, as is readily seen from equation (6) above, to $b_2 = 0$.

$$\frac{1}{y}\frac{dy}{dx} = \frac{x+a}{b_0+b_1 x} = \frac{1}{b_1} + \frac{a-\frac{b_0}{b_1}}{b_0+b_1 x}.$$

Removing the origin to $x = -a$, and writing

$$l = \frac{b_0}{b_1} - a, \qquad \gamma = -\frac{1}{b_1},$$

we obtain the final form

$$y = y_0\left(1 + \frac{x}{l}\right)^{\gamma l} e^{-\gamma x}.$$

The curve is skew, and limited in one direction.

(6) Transition from (1) to (2), $\kappa = 0$, $\beta_1 = 0$. Equation (6) now becomes

$$-\frac{1}{y}\frac{dy}{dx} = \frac{(10\beta_2 - 18)\frac{x}{\sigma}}{\sigma\left\{4\beta_2 + 2(\beta_2 - 3)\frac{x^2}{\sigma^2}\right\}}.$$

Clearly three cases arise, according as β_2 is equal to, greater than, or less than 3.

First consider $\beta_2 = 3$,

$$\frac{1}{y}\frac{dy}{dx} = -\frac{x}{\sigma^2},$$

$$\log y = -\frac{x^2}{2\sigma^2} + \text{constant},$$

$$y = y_0\, e^{-\frac{x^2}{2\sigma^2}}.$$

This is the normal error curve.

(7) Transition from (1) to (2). $\kappa = 0$, $\beta_1 = 0$, $\beta_2 > 3$. Pearson's Type VII.

The differential equation becomes

$$\frac{1}{y}\frac{dy}{dx} = \frac{-(10\beta_2 - 18)\dfrac{x}{\sigma}}{\sigma\left\{4\beta_2 + 2(\beta_2 - 3)\dfrac{x^2}{\sigma^2}\right\}},$$

$$\log y = -m \log\left\{1 + \frac{(\beta_2 - 3)}{2\beta_2}\frac{x^2}{\sigma^2}\right\},$$

$$y = y_0\left(1 + \frac{x^2}{d^2}\right)^{-m}.$$

The curve is symmetrical, bell-shaped, and of unlimited range in both directions.

(8) Transition from (1) to (2). $\kappa = 0$, $\beta_1 = 0$, $\beta_2 < 3$. Pearson's Type II.

It follows as in (7) above that

$$\log y = m \log\left(1 - \frac{x^2}{d^2}\right),$$

$$y = y_0\left(1 - \frac{x^2}{d^2}\right)^{m}.$$

The curve is symmetrical, bell-shaped, and of limited range in both directions.

(9) Special case of (1) above when $5\beta_2 - 6\beta_1 - 9 = 0$, i.e. when the x term in the numerator of equation (6) vanishes. Pearson's Type XII.

Integration of the equation after substitution for β_2 yields

$$y = y_0 \left\{ \frac{\sigma\left(\sqrt{3 + \beta_1} + \sqrt{\beta_1}\right) + x}{\sigma\left(\sqrt{3 + \beta_1} - \sqrt{\beta_1}\right) - x} \right\}^{\sqrt{\beta_1\,(3 + \beta_1)}}.$$

The curve is a twisted **J**-shaped curve.

It will have been noted that type (1) above can be either bell-shaped, **U**-shaped, or **J**-shaped, and that type (3) may be either bell-shaped or **J**-shaped.

The values of β_1 and β_2 which give **J**-shaped curves are restricted within an area bounded by the curve *

$$\beta_1\,(8\beta_2 - 9\beta_1 - 12)\,(\beta_2 + 3)^2 = 4\,(\beta_2 - 3\beta_1)\,(10\beta_2 - 12\beta_1 - 18)^2$$
$$\dots(8).$$

Four special types of distribution can occur along the curve represented by (8). These are shown below as (10) to (13). For fuller details the reader is referred to Pearson's original paper.

(10) $\kappa = \infty$, $\beta_1 = 4$, $\beta_2 = 9$. Pearson's Type X.
This is a special case of V above.
Equation (6) reduces to

$$\frac{1}{y}\frac{dy}{dx} = -\frac{1}{\sigma}, \qquad y = y_0 e^{-\frac{x}{\sigma}}.$$

The curve is **J**-shaped with a finite ordinate at $x = 0$.

(11) $\kappa < 0$, $5\beta_2 - 6\beta_1 - 9 < 0$. Pearson's Type VIII.
The distribution is given by

$$y = y_0 \left(1 + \frac{x}{d}\right)^{-m}.$$

The curve has an infinite ordinate at $x = -d$, and a finite ordinate at $x = 0$.

(12) $\kappa < 0$, $5\beta_2 - 6\beta_1 - 9 > 0$, $2\beta_2 - 3\beta_1 - 6 < 0$. Pearson's Type IX.
The distribution is given by

$$y = y_0 \left(1 + \frac{x}{d}\right)^{m}.$$

* Vide Pearson, *Phil. Trans. Roy. Soc.* 216 A, p. 432.

The curve ranges from $x = -d$ where $y = 0$ to $x = 0$ where $y = y_0$.

(13) $\kappa > 1$, $2\beta_2 - 3\beta_1 - 6 > 0$. Pearson's Type XI.

The distribution is given by

$$y = y_0 \, x^{-m}.$$

The curve is J-shaped and starts at a finite ordinate at $x = b$.

The value of the criterion and of β_1 and β_2 will in all cases determine which of the 13 types should be used to obtain the best representation of the observations under consideration. Types (1), (2) and (3) or Pearson's Types I, IV and VI will cover most of the cases that can arise, but transitional forms will occasionally arise in practice.

For details of the methods best adapted for computing the parameters, the reader is referred to *Frequency Curves and Correlation* by W. Palin Elderton, and *Tables for Statisticians and Biometricians* by Karl Pearson.

65. Use of the series

$$y = A_0 \phi(x) + A_3 \phi'''(x) + A_4 \phi^{iv}(x) + \text{etc.}$$

In these series A_0, A_3, A_4, etc. are constants, and $\phi(x)$ is given by

$$\phi(x) = \frac{1}{\sigma\sqrt{2\pi}} e^{\frac{-(x-b)^2}{2\sigma^2}}, \qquad \phi'''(x) = \frac{d^3}{dx^3}\phi(x), \text{ etc.},$$

σ being the S.D. of the distribution.

The use of this form of frequency curve has been proposed by Thiele[*], Edgeworth[†], and Charlier[‡]. It should be noted that the first term of the series gives the normal error curve. The next term introduces skewness into the curve, while the effect of the third term is symmetrical. Charlier, in the first paper referred to, developed this form of the error law from the hypothesis that an error is made up of a large number of small errors, each of which has its own error law. Edgeworth made considerable use of a functional form which involved only the first two terms in the series.

[*] *Theory of Observations*, London, 1903.

[†] *Camb. Phil. Trans.*, Vol. xx, pp. 36–65, 113–141.

[‡] *Arkiv för Matematik*, Vol. II, Stockholm, 1905, "Über das Fehlergesetz"; *Lunds Meddelanden*, 1906, "Researches into the Theory of Probability."

Charlier fits the curve to the observations by the method of moments. With the notation of the present chapter, this gives

$$b = \mu_1' \qquad \sigma^2 = \mu_2$$

$$A_0 = 1$$

$$3! A_3 = - \mu_3$$

$$4! A_4 = \mu_4 - 3\sigma^4$$

$$5! A_5 = - \mu_5 + 10\sigma^2 \mu_3$$

$$6! A_6 = \mu_6 - 15\sigma^2 \mu_4 + 15\sigma^6.$$

It is thus not difficult to fit a curve to any given series of observations.

66. Other forms of possible frequency curves might be suggested, and with most of these it is not difficult to fit theory to fact. The difficulty lies rather in finding some means of estimating the relative values of the different laws of presumptive errors. We ask in vain for a fixed rule by which the most important and trustworthy forms can be selected. The difficulty is scarcely minimised by the fact that it is often possible to obtain fairly good representations of a given set of observations by two curves whose functional forms differ very widely. Thus Elderton, in his excellent book on *Frequency-Curves and Correlation*, shows that a certain series of observations can be almost equally well represented by the two forms

$$y = y_0 \left(1 + \frac{x^2}{(13 \cdot 39152)^2} \right) e^{-4 \cdot 4504 \tan^{-1} \frac{x}{13 \cdot 39152}},$$

and $y = 4302 [2 \cdot 127818 \, \phi(x) + \cdot 012208 \times (2 \cdot 127818)^4 \phi'''(x)$

$$+ \cdot 007079 \times (2 \cdot 127818)^5 \phi^{iv}(x)],$$

while another series of observations can be represented by the two forms

$$y = 462 \cdot 57 \left(1 - \frac{x^2}{(4 \cdot 543079)^2} \right)^{4 \cdot 141766},$$

and $y = 1244 \cdot 4 [\sigma \phi(x) - \cdot 0081 \, \sigma^4 \phi'''(x) - \cdot 01882 \, \sigma^5 \phi^{iv}(x)],$

where $\sigma^2 = 1 \cdot 829172.$

All that seems at all definite is that with the series

$$y = A_0 \phi(x) + A_3 \phi'''(x) + A_4 \phi^{iv}(x) + \text{etc.}$$

it is difficult to graduate a very skew distribution, or one that rises very rapidly from the axis. For, to do so would involve the use of a large number of terms of the series, so involving the higher moments, whose probable errors are considerably greater than those of the lower moments. In such cases it is better to adhere to Pearson's family of curves. The straightforward use of the criterion will lead to the type of curve which should give the best fit.

CHAPTER X

CORRELATION

67. In the preceding chapters we have only considered frequency distributions due to a single variable, or to more than one *independent* variable. In the present chapter we shall consider the case where the variables are correlated. The method may be illustrated by a simple example. The first two columns of the table on p. 157 below give the orbital period and duration

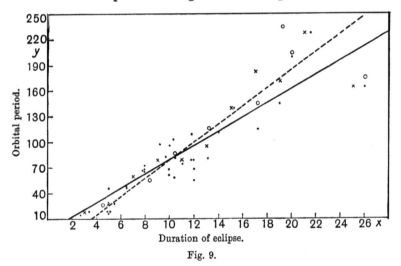

Fig. 9.

of eclipse of 38 Algol stars*. These are represented diagrammatically in figure 9. The horizontal scale (x) represents the duration of eclipse in hours, and the vertical scale (y) represents the orbital period in hours. Each dot in the diagram represents

* Father Stein, *Monthly Notices, R.A.S.*, Vol. LXIX, p. 450.

one star. Such a diagram may be conveniently called a "scatter diagram."

The points in the scatter diagram are roughly grouped about a straight line, showing, as might have been anticipated *a priori*, that the duration of eclipse generally increases as the orbital period increases. The total range of duration of eclipse is sub-divided into ranges 2^h—4^h, 4^h—6^h, etc. The mean orbital period for all the stars in each of these intervals is calculated, and repre-sented in the diagram by means of a × placed at the middle of the range, e.g. the mean orbital period of stars whose duration of eclipse is between 10^h and 12^h is 79^h, and this mean is represented in the diagram by a × at 11^h, 79^h. Similarly the mean duration of eclipse for stars whose orbital periods are between 10^h and 40^h, between 40^h and 70^h, etc., are represented in the diagram by small circles.

A straight line is drawn to fit the series of crosses as accurately as possible. In this particular instance the crosses do not lie accurately on the straight line, but lie irregularly on either side of it. This straight line is called *the line of regression of y on x*. Its ordinate gives the mean value of *y*, which we may expect to find associated with a given value of *x*. Similarly the straight line which fits most accurately the series of small circles is called *the line of regression of x on y*. Its abscissa gives the mean value of *x* corresponding to a given value of *y*. It should be noted that the two lines of regression do not coincide, though the angle between them is small. When the means lie fairly accurately on a straight line, the regression is said to be *linear*. But it may happen that the means lie, not on a straight line, but on a well-defined curve. The curve is called the *curve of regression*, and the regression is then defined as *non-linear*. Such a curve would be obtained if a number of observations of the volume (v) and the pressure (p) of a given mass of gas at a constant temperature were represented by a scatter diagram. The points in the scatter diagram would lie closely about a rectangular hyperbola, whose equation is $pv = $ constant. Even in cases where there is a clearly defined non-linear curve of regression, the straight lines which fit most closely the series of means are called the lines of regression.

Generally speaking, it is only in dealing with isolated physical phenomena, in which the conditions of observation can be completely controlled, that we shall find a clearly defined functional relation between the two variables considered. When the factors which vary are complex and not controllable by the observer, the curve of regression does not as a rule indicate a simple functional relationship between the two characters considered. The complete problem of the statistician, which is to find formulae which will represent with sufficient accuracy the form of the curve of regression, is not in general capable of solution. But since the majority of the problems of the practical statistician relate solely to averages, it is sufficient in many cases to be able to state whether, on an average, there is a tendency for high values of one of the characters to be associated with high values (or low values) of the other. If possible it is also desirable to find how great a divergence of one character from its mean value is associated with a unit divergence of the other from its mean value; and also how closely this relation is usually fulfilled. These questions can be largely decided by fitting a straight line to the series of means obtained as in figure 9, i.e. by drawing the lines of regression.

68. As an alternative to the scatter diagram, we might represent the frequency distribution of two variables by means of a table of double entry, or a *contingency table*. The following contingency table represents the material in the scatter diagram of figure 9.

Each row in this table gives the frequency distribution of the duration of eclipse for a given range of orbital period, while each column gives the frequency distribution of the orbital period for a given range of duration of eclipse. As the columns and rows are only distinguished by the accidental circumstance that one runs vertically and the other horizontally, the word "array" is used to denote either a row or a column.

The choice of class-intervals is to a large extent arbitrary, and in general any interval which happens to be convenient may be selected. When this choice has been made, the contingency table can be completed in a number of ways. If a scatter diagram such as that of figure 9 has been made, the class-intervals may be ruled in the diagram, and the number of dots in each compartment may then be counted. If such a scatter diagram is not available, a form such as that shown below should be ruled on a large sheet of paper, with the class-intervals headed as in the final table. Each observation can be represented in this table by a cross in the corresponding compartment. The sum

Duration of eclipse in hours.

	2—	4—	6—	8—	10—	12—	14—	16—	18—	20—	22—	24—	Total
10—	2	6	—	—	—	—	—	—	—	—	—	—	8
40—	—	1	3	2	3	2	—	—	—	—	—	—	11
70—	—	—	—	3	4	2	—	—	—	—	—	—	9
100—	—	—	—	—	2	—	1	1	—	—	—	—	4
130—	—	—	—	—	—	—	1	—	1	—	—	—	2
160—	—	—	—	—	—	—	—	—	—	—	—	1	1
190—	—	—	—	—	—	—	—	—	—	1	—	—	1
220—	—	—	—	—	—	—	—	1	—	1	—	—	2
Total	2	7	3	5	9	4	2	2	1	2	—	1	38

(left axis label: Orbital period in hours.)

of the crosses in a compartment is then entered in the corresponding compartment of the final table. It should be noted that any compartment may contain halves and even quarters of a frequency. For if an observation falls exactly on the dividing line between two compartments, it is counted as a half in each of those compartments; and if an observation falls exactly at the common angular point of four compartments, it is counted as a quarter in each of the four compartments.

When there is a functional relation between the two characters considered, as will often happen in the problems of the physicist or chemist, there will be entries in only a few compartments of the contingency table. The way in which the entries are grouped will afford some idea of the relationship between the two variable characters. Thus in the contingency table above, the entries run diagonally across the table, showing distinct correlation between the two characters considered.

When the means represented by crosses (or circles) in the scatter diagram (see figure 9) lie accurately on a straight line, the two variables considered are connected by a linear relation, and are said to be completely correlated. In practice, however, this seldom occurs, and the means lie irregularly about a straight line. The straight line which best fits the series of means might be drawn by a simple graphical method, say by means of a stretched thread moved about until as many of the means lie on one side as on the other. But such a method would generally allow of our drawing a number of straight lines, any one of which would apparently be as good a fit as any other. It is therefore necessary to adopt some standard method of drawing what shall be regarded as the best-fitting straight line. The method commonly adopted is based on the theory of least squares, but it should be remembered that this step is arbitrary, and that other methods might be suggested which would yield equally good results. The method of moments yields precisely the same results as the least squares method*.

69. The total range of x-variation is divided into a con-

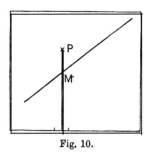

Fig. 10.

venient number of intervals. Let there be n_x observations, i.e. n_x y's, in any one of these intervals, and let the mean of these y's be represented by the ordinate of the point P in figure 10. A number of points such as P will be represented in the completed figure, each point corresponding to the mean of all the y's in one x-interval. The curve of regression passes through all these points. We now have to draw the straight

* Elderton, *Frequency-Curves and Correlation*, p. 114.

line which shall best fit all these points, or, in other words, we
have to find the equation of the line of regression of y on x.

Referred to axes through the means of x and y, let the
equation of the line of regression be

$$Y = a + bX.$$

Let the ordinate of P meet the line of regression at the point
M. Then the usual least squares method of adjustment is to
make $\Sigma n_x PM^2$ a minimum. Let the ordinate of P be y_x, and
of M, Y. Then we have to make

$$V = \Sigma n_x (Y - y_x)^2$$

a minimum, where the summation extends to all the x-intervals.

Since $Y = a + bX$ for the point M,

$$V = \Sigma n_x \{(a + bX) - y_x\}^2$$
$$= \Sigma n_x \{a^2 + b^2 X^2 + 2abX + y_x^2 - 2ay_x - 2bX y_x\}.$$

But since the origin is at the mean of x and y,

$$\Sigma n_x y_x = 0, \qquad \Sigma n_x X = 0,$$

and therefore

$$V = \Sigma n_x \{a^2 + b^2 X^2 + y_x^2 - 2bX y_x\}.$$

This last form of V clearly demands $a = 0$ for its minimum
value. Differentiating with respect to b,

$$2b\Sigma n_x X^2 - 2\Sigma n_x X y_x = 0$$

and

$$b = \frac{\Sigma n_x X y_x}{\Sigma n_x X^2} = \frac{\Sigma xy}{\Sigma x^2},$$

where, in the last fractional form, the summation extends to all
pairs of associated values of x and y. If σ_1, σ_2 be the standard
deviations of x and y respectively,

$$\Sigma x^2 = N\sigma_1^2, \qquad \Sigma y^2 = N\sigma_2^2.$$

Let $\qquad \Sigma xy = N r \sigma_1 \sigma_2$, or $\quad r = \dfrac{\Sigma xy}{\sqrt{\Sigma x^2 . \Sigma y^2}}.$

Then $b = r\dfrac{\sigma_2}{\sigma_1}$, and the equation of the line of regression of
y on x is

$$y = r\frac{\sigma_2}{\sigma_1} x = \frac{\Sigma xy}{\Sigma x^2} . x,$$

or

$$\frac{y}{\sigma_2} = r\frac{x}{\sigma_1}.$$

Similarly the line of regression of x on y may be shown to have the equation

$$x = r\frac{\sigma_1}{\sigma_2}y = \frac{\Sigma xy}{\Sigma y^2}\cdot y,$$

or

$$\frac{x}{\sigma_1} = r\frac{y}{\sigma_2}.$$

If the measures x, y be referred to some other zero than the mean, and $\bar{x}$, $\bar{y}$ be their mean values, the equations of the lines of regression given above are changed into

$$y - \bar{y} = r\frac{\sigma_2}{\sigma_1}(x - \bar{x}),$$

$$x - \bar{x} = r\frac{\sigma_1}{\sigma_2}(y - \bar{y}).$$

In these equations r is a quantity defined by the equation

$$\Sigma xy = Nr\sigma_1\sigma_2, \quad \text{or} \quad r = \frac{\Sigma xy}{N\sigma_1\sigma_2} = \frac{\Sigma xy}{\sqrt{\Sigma x^2 \Sigma y^2}}.$$

This quantity r is called *the coefficient of correlation*.

70. Returning to the function V defined above, we may now write

$$V = \Sigma n_x (bX - y_x)^2,$$

where the summation extends to all the x-arrays,

or

$$V = \Sigma (bx - y)^2 = \Sigma \left(r\frac{\sigma_2}{\sigma_1}x - y\right)^2,$$

where the summation extends to all the pairs of associated deviations.

Hence

$$V = \frac{r^2\sigma_2^2}{\sigma_1^2}\Sigma x^2 - \frac{2r\sigma_2}{\sigma_1}\Sigma xy + \Sigma y^2$$

$$= N\left\{\frac{r^2\sigma_2^2}{\sigma_1^2}\sigma_1^2 - 2r^2\sigma_2^2 + \sigma_2^2\right\}$$

$$= N\sigma_2^2(1 - r^2).$$

If s_2 be the M.S.E. due to taking the value of y given by the line of regression

$$y = \frac{r\sigma_2}{\sigma_1}x$$

instead of the measured deviation of y, then

$$V = Ns_2^2 = N\sigma_2^2(1 - r^2),$$

and

$$s_2 = \sigma_2(1 - r^2)^{\frac{1}{2}}.$$

Similarly if s_1 be the M.S.E. of x derived from the equation

$$x = \frac{r\sigma_1}{\sigma_2} y,$$

then $s_1 = \sigma_1(1 - r^2)^{\frac{1}{2}}.$

If $r = 1$, then $V = 0$, and $s_1 = s_2 = 0$. And since

$$V = \Sigma (bx - y)^2$$

it follows that when $r = 1,$

$$bx - y = 0$$

for each pair of associated deviations x, y. In other words, a linear relation

$$\frac{x}{\sigma_1} - \frac{y}{\sigma_2} = 0$$

is then rigorously satisfied by all pairs of values of x, y.

If the coefficient r only differs slightly from unity, the points in the scatter diagram are closely grouped about a straight line. The two lines of regression (which coincide when $r = 1$) are then inclined to one another at a small angle.

If r be small the angle between the lines of regression is large, and when $r = 0$, the lines are at right angles, their equations being $y = 0$, $x = 0$. In the case where r is small, the M.S.E. of y caused by our adopting the linear relation $y = \frac{r\sigma_2}{\sigma_1} x$ between the two variables instead of using the original observations, defined above as s_2, is nearly as great as σ_2. Or in other words, if we want to find the value of y corresponding to a given value of x, the value $\frac{r\sigma_2}{\sigma_1} x$ is only a very slight improvement on the mean value of all the y's in the case where r is small.

When r is small, there is only very slight correlation between the two variable characters considered, and it seems doubtful whether any serious meaning can be attached to values of r which are less than ·5.

When $r = 0$, there is apparently no correlation between x and y, and the lines of regression are at right angles to one another. In the next diagram (fig. 11), however, is shown an extreme case where $r = 0$, while the variables are connected by a clearly marked relation. It is therefore not safe to assume a complete absence of correlation in cases where the coefficient r is very small. The evaluation of the correlation ratio, which will be discussed later, affords a better test of correlation in such cases.

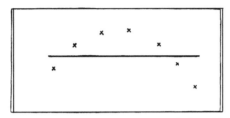

Fig. 11.   Curve of regression of y on x, together with the corresponding line of regression.

71. Evaluation of r for the material of figure 9.

In the following table we shall evaluate r for the material which has been represented diagrammatically in figure 9. The first column gives the orbital period in hours, and the second column the duration of eclipse in hours, for 38 stars. The means of these columns are evaluated. In the third and fourth columns are given the deviations from these means of the orbital period and duration of eclipse of each star. These are the quantities which we have called y, and x, respectively, in the preceding discussion. The remainder of the table is self-explanatory*.

$$r = \frac{9787 \cdot 7}{\sqrt{117470 \times 1050}} = \cdot 881,$$

$$\frac{\Sigma xy}{\Sigma x^2} = \frac{9787 \cdot 7}{1050} = 9 \cdot 32.$$

$$\frac{\Sigma xy}{\Sigma y^2} = \frac{9787 \cdot 7}{117470} = \cdot 083.$$

The corresponding lines of regression are

$$Y - 82 \cdot 7 = 9 \cdot 32 \, (X - 10 \cdot 4),$$

or $$Y = 9 \cdot 32 X - 14 \cdot 2 \quad \dots\dots\dots\dots(A),$$

and $$X - 10 \cdot 4 = \cdot 083 \, (Y - 82 \cdot 7)$$

or $$X = \cdot 083 Y + 3 \cdot 6 \quad \dots\dots\dots\dots(B).$$

* See also § 79 below.

Orbital period	Duration of eclipse	y	x	y^2	x^2	xy +	xy −
13·2	2·7	− 69·5	− 7·7	4830	59·3	535·2	
15·1	3·5	− 67·6	− 6·9	4570	47·6	466·4	
20·1	5·1	− 62·6	− 5·3	3919	28·1	341·8	
21·4	5·0	− 61·3	− 5·4	3758	29·2	331·0	
21·6	5·0	− 61·1	− 5·4	3733	29·2	329·9	
27·3	5·0	− 55·4	− 5·4	3069	29·2	299·2	
28·7	5·5	− 54·0	− 4·9	2916	24·0	264·6	
32·6	4·7	− 50·1	− 5·7	2510	32·5	285·6	
45·5	5·0	− 37·2	− 5·4	1384	29·2	200·9	
45·8	6·5	− 36·9	− 3·9	1362	15·2	143·9	
47·4	6·5	− 35·3	− 3·9	1246	15·2	137·7	
55·8	12·0	− 26·9	1·6	724	2·6		43·0
58·0	10·4	− 24·7	0	610	0	0	
59·8	10·0	− 22·9	− ·4	524	·2	9·2	
65·4	8·0	− 17·3	− 2·4	299	5·8	41·5	
66·4	8·0	− 16·3	− 2·4	266	5·8	39·1	
66·4	7·9	− 16·3	− 2·5	266	6·3	40·8	
68·8	12·0	− 13·9	1·6	193	2·6		22·2
68·8	10·0	− 13·9	− ·4	193	·2	5·6	
71·9	8·0	− 10·8	− 2·4	117	5·8	25·9	
73·3	11·1	− 9·4	·7	88	·5		6·6
79·3	12·0	− 3·4	1·6	12	2·6		5·4
79·6	11·8	− 3·1	1·4	10	2·0		5·3
81·1	13·1	− 1·6	2·7	3	7·3		4·3
82·8	10·5	·1	·1	0	0	0	
82·8	9·7	·1	− ·7	0	·5		·1
94·9	10·0	12·2	− ·4	149	·2		4·9
95·8	9·2	13·1	− 1·2	172	1·5		15·7
106·2	10·3	23·5	− ·1	552	0		2·4
109·5	11·8	26·8	1·4	718	2·0	37·5	
110·4	14·0	27·7	3·6	767	13·0	99·7	
115·3	17·2	32·6	6·8	1063	46·2	221·7	
142·4	15·2	59·7	4·8	3564	23·0	286·6	
144·1	19·0	61·4	8·6	3770	74·0	528·0	
164·7	26·0	82·0	15·6	6724	243·4	1279·2	
202·3	20·0	119·6	9·6	14303	92·2	1148·2	
227·6	21·5	144·9	11·1	20996	123·2	1608·4	
250·3	17·5	167·6	7·1	28090	50·4	1190·0	
Sums: 3142·4	400·7			117470	1050·0	9787·7	
Means: 82·7	10·4			3091·3	27·63	257·54	

$$\sigma_2 = 55\cdot6 \quad \sigma_1 = 5\cdot26$$

The two straight lines are represented in figure 9 by the continuous and dotted lines respectively.

The mean square error in y due to assuming the relation A is

$$55\cdot6\sqrt{1 - (\cdot881)^2}\ \text{hours} = 26\cdot4\ \text{hours},$$

and the mean square error in x due to assuming the relation B is

$$5{\cdot}26\sqrt{1-({\cdot}881)^2}\text{ hours} = 2{\cdot}5\text{ hours}.$$

72. Calculation of r from the Contingency Table.

When the number of observations is large, the method of calculating r used in the preceding table becomes extremely laborious. It is then better to calculate r from the contingency table. To illustrate the method, we shall evaluate r from the contingency table of page 151.

Duration of eclipse.

	2—	4—	6—	8—	10—	12—	14—	16—	18—	20—	22—	24—	Totals
10—	$_8$2	$_6$6											8
40—		$_3$1	$_2$3	$_1$2	$_0$3	$_1$2							11
70—				$_0$3	$_0$4	$_0$2							9
100—					$_0$2		$_2$1	$_3$1					4
130—							$_4$1		$_8$1				2
160—												$_{21}$1	1
190—										$_{20}$1			1
220—								$_{15}$1		$_{25}$1			2
Totals	2	7	3	5	9	4	2	2	1	2	—	1	38

(left side vertical label: Orbital period.)

The above table is a repetition of the table of page 151, with the arrays which contain the means of the two variables marked by thick lines. The midpoints of these two arrays are taken as zeros for the coordinates, and the class-intervals are taken as units in terms of which the S.D.'s of x and y are to be expressed. The following tables give the calculation.

Each of the tables (a) and (b) gives the calculation of the mean and of the S.D. for one of the variables. For still greater accuracy, we might apply Sheppard's corrections to the values of σ_1 and σ_2.

(a) For x.

Sum	Factor	1st product	2nd product
2	−4	−8	32
7	−3	−21	63
3	−2	−6	12
5	−1	−5	5
9	0	0	0
4	1	4	4
2	2	4	8
2	3	6	18
1	4	4	16
2	5	10	50
0	6	0	0
1	7	7	49
38		−5	257
Means ..	... ...	−·13	6·763

$$\sigma_1^2 = 6\cdot763 - (\cdot13)^2 = 6\cdot746. \qquad \sigma_1 = 2\cdot6.$$

(b) For y.

Sum	Factor	1st product	2nd product
8	−2	−16	32
11	−1	−11	11
9	0	0	0
4	1	4	4
2	2	4	8
1	3	3	9
1	4	4	16
2	5	10	50
38		−2	130
Means ...	... ...	−·05	3·43

$$\sigma_2^2 = 3\cdot43 - (\cdot05)^2 = 3\cdot427. \qquad \sigma_2 = 1\cdot85.$$

These tables yield almost exactly the same values of σ_1 and σ_2 as were derived from the ungrouped material of observation.

Thus $\qquad \sigma_1 = 2\cdot6$ units = 5·2 hours,

$\sigma_2 = 1\cdot85$ units = 1·85 × 30 hours = 55·5 hours.

In order to complete the calculation of r, we must calculate the product-moment Σxy. The factors corresponding to xy are the small figures given in the lower left-hand corners of the compartments of the table. In this particular example the sum of all the moments from any one quadrant can easily be evaluated. The result may be represented thus :

$$
\begin{array}{c|c}
+63 & -2 \\
\hline
0 & +98
\end{array}
\qquad \text{Total 159.}
$$

The product-moment thus obtained is not about the means of the two variables, and must be corrected.

$$r\sigma_1\sigma_2 + {\cdot}13 \times {\cdot}05 = \frac{159}{38} = 4{\cdot}19,$$

$$r\sigma_1\sigma_2 = 4{\cdot}183,$$

$$\therefore \ r = \frac{4{\cdot}183}{2{\cdot}6 \times 1{\cdot}85} = {\cdot}870.$$

Thus the value of r yielded by this method is ·870 as compared with ·881 obtained by working with the original ungrouped material.

When the number of observations represented in the contingency table is large, the product-moment cannot be obtained quite as simply as was done above. It may then be advisable to write in one column all the factors which are effective, writing the corresponding frequencies, with due regard to sign, in a second column. The remainder of the calculation is then easy.

73. The Correlation Ratio.

If the curve of regression be not linear, r cannot be regarded as a satisfactory measure of the amount of correlation between the two characters considered. We then have to find a method which will decide whether the observations are clustered closely about the curve of regression, or are widely scattered. The obvious method is to evaluate the S.D. of each array.

Let s_{ax} be the S.D. of any x-array. Then once the quantities

s_{ax} have been evaluated for all the arrays, we can tell accurately how closely the individual points in the scatter diagram are clustered about the curve of regression. A curve representing $\dfrac{s_{ax}}{\sigma_1}$ is called a *scedastic curve*. The mean ordinate of such a curve is a measure of the closeness with which a given value of x yields a definite value of y. In practice, however, it is more convenient to evaluate the S.D. of the weighted S.D.'s of the separate x-arrays. Let this S.D. be σ_{ax}.

Then
$$\sigma^2{}_{ax} = \frac{\Sigma n s^2{}_{ax}}{N}.$$

It is found that $\sigma^2{}_{ax}$ can be calculated without going through the process of evaluating s_{ax} for each x-array.

Let M_y denote the mean of all y's, m_y the mean of any x-array [*] whose S.D. is s_{ax}. Summing for one array, we find
$$\Sigma (y - M_y)^2 = \Sigma (y - m_y)^2 + n (M_y - m_y)^2$$
$$= n s^2{}_{ax} + n (M_y - m_y)^2.$$
Summing for all arrays, we find
$$N\sigma_2{}^2 = N\sigma^2{}_{ax} + N\sigma^2{}_{my},$$
where σ_{my} is the S.D. of the weighted means of the separate x-arrays.
$$\therefore \quad \sigma^2{}_{ax} = \sigma_2{}^2 - \sigma^2{}_{my} = \sigma_2{}^2 (1 - \eta^2{}_{yx}) \text{ say,}$$
where
$$\eta_{yx} = \frac{\sigma_{my}}{\sigma_2}.$$

The quantity η_{yx} thus defined is called the *correlation ratio* of y on x. There will naturally be a second correlation ratio, η_{xy}, which is the correlation ratio of x on y. Its value is $\dfrac{\sigma_{mx}}{\sigma_1}$.

The mean square of the distance between the curve of regression and the line of regression $y = r \dfrac{\sigma_2}{\sigma_1} x$ is
$$\frac{1}{N} \Sigma n_y \left(\overline{y}_x - r \frac{\sigma_2}{\sigma_1} x \right)^2 = \frac{\Sigma n_y \overline{y}_x{}^2}{N} + \frac{r^2 \sigma_2{}^2}{\sigma_1{}^2} \frac{\Sigma n_y x^2}{N} - 2r \frac{\sigma_2}{\sigma_1} \frac{\Sigma n_y x \overline{y}_x}{N}$$
$$= \sigma^2{}_{my} + \frac{r^2 \sigma_2{}^2}{\sigma_1{}^2} \sigma_1{}^2 - 2r \frac{\sigma_2}{\sigma_1} r \sigma_1 \sigma_2$$
$$= \sigma^2{}_{my} - r^2 \sigma_2{}^2$$
$$= \sigma_2{}^2 (\eta^2{}_{yx} - r^2).$$

[*] An x-array is an array of y's.

Thus $\eta^2 - r^2$ is a measure of the deviation from linearity of the curve of regression. In accurate work it is advisable to calculate η as well as r, since η is a better measure of causal relation than r, and $\eta^2 - r^2$ affords a measure of the linearity of the regression. It should be noted that η is always greater than r, except when the regression is accurately linear, and in this case $\eta = r$. And conversely, if $\eta = r$, within the limits of random sampling, the regression is linear.

It is clear from the equation

$$1 - \eta^2{}_{yx} = \frac{\sigma^2{}_{ax}}{\sigma_2{}^2}$$

that $|\eta|$ must lie between 0 and 1, and since η is the ratio of two S.D.'s it must be positive. The correlation ratio therefore lies between 0 and 1 in all cases. When the correlation is complete η is unity, and when there is no correlation between the variables, η is zero. When there is a considerable amount of correlation between the variables, η is large, and when the variables are only slightly correlated η is small. But the correlation ratio only affords a satisfactory test when the number of observations is sufficiently great to permit of the formation of a grouped contingency table.

74. Probable Errors.

As all the quantities σ, r, η, etc., mentioned above are in general calculated from a sample of the total population which bears the characters considered, it is necessary to consider the probable errors of all these quantities. Remembering that

$$\sigma = \mu \sqrt{\frac{n-1}{n}},$$

we may write

P.E. of mean $= \cdot6745 \dfrac{\mu}{\sqrt{n}} = \cdot6745 \dfrac{\sigma}{\sqrt{n}}$ (approximately),

P.E. of $\sigma = \cdot6745 \dfrac{\sigma}{\sqrt{2(n-1)}} = \cdot6745 \dfrac{\sigma}{\sqrt{2n}}$ (approximately).

The P.E. of r may be taken to be

$$\cdot6745 \frac{1-r^2}{\sqrt{n}},$$

in cases where the frequency distribution is approximately normal, and n is large.

When the regression is linear, or nearly linear, the P.E. of η is

$$\cdot 6745 \, \frac{1 - \eta^2}{\sqrt{n}} \, .$$

These formulae are based upon certain definite assumptions concerning the relations between the variables considered. It can be shown that if the variations of x and y from their mean values are due to a number of independent sources of error, some of which are common to both, while each of the elementary errors follows a normal law of distribution, then the number of cases in which x lies between x and $x + dx$, while y lies between y and $y + dy$, will be $f(x, y)\,dx\,dy$, where

$$f(x, y) = A\,e^{-\frac{1}{2(1-r^2)}\left(\frac{x^2}{\sigma_1^2} + \frac{y^2}{\sigma_2^2} - \frac{2xyr}{\sigma_1\sigma_2}\right)} \, .$$

It will be noted that this law of distribution requires that both x and y, when considered independently, should follow a normal law of distribution.

The P.E.'s of r and η given above are calculated by the use of the formula here given for $f(x, y)$. The methods of correlation are frequently applied to variables which do not follow a normal law of variation, and the probable errors of r and η are still calculated by the use of the formulae given above. It must be remembered that in such cases the formulae cannot yield more than a rough approximation to the values of the probable errors.

75. To calculate η for the material of the table on page 153.

The following is a rough estimate of the value of η. It is necessary to evaluate m_y for each x-array. For the purpose of a rough calculation, it is sufficient to assume the mean value of y for the observations in any one compartment to be the central value for that compartment. The value of m_y is then easily evaluated. The mean value of all the y's has already been shown to be 82·7. This is the quantity called M_y.

x	n	m_y	$M_y - m_y$	$(M_y - m_y)^2$	$n(M_y - m_y)^2$
2— 4	2	25·0	− 57·7	3329	6658
4— 6	7	29·3	− 53·4	2851	19957
6— 8	3	55·0	− 27·7	765	2295
8—10	5	73·0	− 9·7	94	470
10—12	9	81·7	− 1·0	1	19
12—14	4	70·0	− 12·7	161	644
14—16	2	130·0	47·3	2237	4474
16—18	2	175·0	92·3	8519	17038
18—20	1	145·0	62·3	3581	3581
20—22	2	220·0	137·3	18851	37702
22—24	—	—	—	—	—
24—26	1	175·0	92·3	8519	8519
	38				101357

$$\sigma^2_{my} = \frac{101357}{38} = 2667 \qquad \sigma_{my} = 51\cdot6.$$

$$\therefore \ \eta = \frac{\sigma_{my}}{\sigma_2} = \frac{51\cdot6}{55\cdot6} = \cdot928.$$

The value of r was found to be ·881.

76. Spearman's Formulae. The Method of Ranks*.

In the method of ranks the actual measurements are replaced by the numbers representing the ranks of measurements when they are arranged in order of magnitude. Spearman applied the method to the consideration of the correlation of psychical performances in individuals. If N individuals be considered, and ν_1, ν_2 be the orders of merit of any one individual in the two series of measurements, the product-moment reduces to

$$1 - \frac{6\Sigma\,(\nu_1 - \nu_2)^2}{N\,(N^2 - 1)} = \rho.$$

Spearman also suggested a simpler formula, for the product-moment,

$$1 - \frac{\Sigma d}{\frac{1}{6}\,(N^2 - 1)} = R,$$

where Σd is the sum of gains in rank (or the sum of positive differences) of the second series on the first.

The method of deduction of these formulae assumes that the

* *Amer. Jour. Psych.*, Vol. xv, 1904; *Brit. Jour. Psych.*, Vol. ii, part 1.

frequency distribution can be represented by a rectangle. Pearson*
has shown that if the frequency distribution is Gaussian, the
quantities ρ and R are connected with r by the relations

$$r = 2 \sin\left(\frac{\pi}{6}\rho\right), \qquad r = 2\cos\frac{\pi}{3}(1 - R) - 1.$$

When the frequency distribution is not Gaussian, it is not
clear how r is to be deduced from ρ or R. For this reason, the
above formulae are not to be recommended for use in accurate
work. A further disadvantage is that the method does not yield
the values of the standard deviations. The formulae may be
useful however in rough work involving only a small number
of observations.

In order to compare the results yielded by Spearman's
formulae with the result deduced from the product-moment
formula, we shall apply these formulae to the example of
page 157. The stars are arranged in order of magnitude of
orbital period in the table of page 157. The positions of these
stars in a table arranged in order of magnitude of eclipse-
duration are given in the second column below.

ν_1	ν_2	d	d^2	ν_1	ν_2	d	d^2
1	1	0	0	20	13	7	49
2	2	0	0	21	24	−3	9
3	8	−5	25	22	27	−5	25
4	4	0	0	23	25	−2	4
5	4	1	1	24	30	−6	36
6	4	2	4	25	23	2	4
7	9	−2	4	26	17	9	81
8	3	5	25	27	18	9	81
9	4	5	25	28	16	12	144
10	10	0	0	29	21	8	64
11	10	1	1	30	25	5	25
12	27	−15	225	31	31	0	0
13	22	−9	81	32	33	−1	1
14	18	−4	16	33	32	1	1
15	13	2	4	34	35	−1	1
16	13	3	9	35	38	−3	9
17	12	5	25	36	36	0	0
18	27	−9	81	37	37	0	0
19	18	1	1	38	34	4	16

* *Drapers' Company Research Memoirs*, Biometric Series IV.

$$\Sigma d = 17. \qquad \Sigma d^2 = 1077.$$

$$\therefore R = 1 - \frac{6 \cdot 17}{38^2 - 1} = 1 - \cdot 071 = \cdot 929,$$

$$\rho = 1 - \frac{6 \cdot 1077}{38\,(38^2 - 1)} = \cdot 882.$$

Using the formulae given above, we find

$$r = 2 \sin\left(\frac{\pi}{6} \times \cdot 882\right) = \cdot 891,$$

$$r = 2 \cos \frac{\pi}{3}\,(\cdot 071) - 1 = \cdot 994.$$

The value derived for r by the use of the product-moment formula is ·881.

77. The Method of Contingency.

The methods discussed above are only applicable in cases where the characters considered are capable of quantitative measurement. Such characters would be length, time, stellar magnitude, wages, etc.   When the grouping in the contingency table is purely classificatory, the different classes bear no quantitative relation to each other.   Thus if stars be grouped according to a colour scale, the scale cannot be regarded as an arithmetical scale.   Except for this difference in the nature of the scale, tables which represent the distribution of two characters, of which one or both may be purely qualitative, are of precisely the same form as the contingency table given on page 151.   The following is an example of such a contingency table*, representing the distribution of 1360 stars according to spectral type and colour.   The groupings a, b, c, d, e, f are according to a scale of colour, where

stars of class a are white,
 „   „ b „ white with faint tinge of colour,
 „   „ c „ very pale yellow,
 „   „ d „ pale yellow,
 „   „ e „ full yellow,
 „   „ f „ ruddy.

* W. S. Franks, "On the Relation between Star Colours and Spectra," *Monthly Notices*, R.A.S., Vol. LXVII, p. 541, Table IV.

Spectral Type	a	b	c	d	e	f	Total
Helium Stars	125 (61·38)	146 (103·05)	8 (44·37)	3 (37·74)	0 (33·18)	0 (2·28)	282
Hydrogen Stars	168 (82·05)	195 (137·77)	14 (59·32)	0 (50·45)	0 (44·35)	0 (3·05)	377
a Carinae Type	3 (29·82)	97 (50·07)	23 (21·56)	8 (18·33)	6 (16·12)	0 (1·11)	137
Solar Stars	0 (39·18)	41 (65·78)	77 (28·32)	33 (24·09)	29 (21·18)	0 (1·46)	180
Arcturus Type	0 (52·45)	15 (88·07)	86 (37·92)	77 (32·25)	63 (28·35)	0 (1·95)	241
Aldebaran Type	0 (16·32)	0 (27·41)	4 (11·80)	22 (10·04)	43 (8·82)	6 (·61)	75
Betelgeuse	0 (14·80)	3 (24·85)	2 (10·70)	39 (9·10)	19 (8·00)	5 (·55)	68
Total	296	497	214	182	160	11	1360

Leaving out of consideration for the moment the figures in brackets in each compartment, we have a contingency table resembling that of page 151. Even a casual examination of this table shows that there is a tendency for stars low down in the scale of spectra to have colours which are nearer to red than those of stars higher in the scale of spectra ; i.e. there is some degree of correlation between colour and spectral type of stars.

The last row in the table gives the distribution according to colour, and the last column the distribution according to spectral type, of all the stars considered. If spectral type and colour were independent factors, the number of stars in any compartment of the table could be evaluated from the totals in the last row and the last column ; e.g. if colour were independent of spectral type, the number of stars of colour c and of solar type would be

$$\frac{214}{1360} \times 180, \text{ or } 28 \cdot 32.$$

The numbers which should occur in the other compartments can be evaluated in the same way. These numbers are entered in the compartments of the table, in brackets. These bracketed figures give the distribution to be expected if there were no correlation between spectral type and colour. The differences between the two sets of figures are due to correlation between the two characters.

It is customary to calculate the amount of correlation in the following way:

Form the difference between the bracketed and unbracketed figures in each compartment; square this difference and divide the result by the corresponding bracketed figure. Then add together all the numbers obtained in this manner, and divide the sum by the total number of stars. This result is called the *total mean square contingency*, and is denoted by ϕ^2.

For the example shown in the above table,

$$1360\,\phi^2 = \frac{(125 - 61\cdot38)^2}{61\cdot38} + \frac{(146 - 103\cdot05)^2}{103\cdot05} + \text{etc.},$$

there being 42 terms in all;

$$\phi^2 = 1\cdot021.$$

The function c defined by

$$c = \sqrt{\frac{\phi^2}{1 + \phi^2}}$$

is called the *coefficient of mean square contingency*. The quantity c is of the same nature as the coefficient of correlation r. When the frequency distribution is normal, it can be shown that c calculated as above is equal to r. It is clear from its definition that c must lie between -1 and $+1$. When there is no correlation between the two characters considered,

$$\phi = 0 \quad \text{and} \quad c = 0.$$

Small values of ϕ and c indicate a small amount of correlation.

For the table given above

$$c = \cdot71,$$

a value which may be taken to indicate a close relationship between spectral type and colour.

For a full treatment of the subject of contingency the reader is referred to the original memoir of Prof. Karl Pearson, "On the Theory of Contingency and its Relation to Association and Normal Correlation," *Drapers' Company Research Memoirs*, Biometric Series I, 1904, Dulau and Co.

Example. Evaluate the s. d. of a linear function of a number of correlated variables.

Let

$$F = ax + by + cz + \ldots$$

be the function.

Then $F^2 = a^2x^2 + b^2y^2 + c^2z^2 + 2abxy + 2acxz + 2bcyz + \ldots.$

It follows that

$$\sigma_F^2 = a^2\sigma_x^2 + b^2\sigma_y^2 + c^2\sigma_z^2 + 2abr_{xy}\sigma_x\sigma_y + 2acr_{xz}\sigma_x\sigma_z + 2bcr_{yz}\sigma_y\sigma_z + \ldots.$$

A special case is $\sigma^2_{x-y} = \sigma_x^2 + \sigma_y^2 - 2r_{xy}\sigma_x\sigma_y.$

This affords a method of evaluating r_{xy} by evaluating

$$\sigma_x, \quad \sigma_y, \quad \sigma_{x-y}.$$

78. The Meaning of the Correlation Coefficient.

The meaning which we shall assign to the correlation coefficient will to some extent depend upon the point of view from which we approach the question. Referring back to § 70 we see that if the coefficient of correlation r between two variables is known, the value of x corresponding to any given value of y can be given with a M.S.E. $s_2 = \sigma_2\sqrt{1 - r^2}$, σ_2 being the M.S.E. of all the y's. Our knowledge of the correlation r thus enables us to reduce the M.S.E. of our estimate of the value of y from σ_2 to $\sigma_2\sqrt{1 - r^2}$. From this point of view it would appear preferable to measure correlation by $\sqrt{1 - r^2}$ rather than by r.

It is customary, however, to regard r as being in some way a measure of the number of common causes which underlie the variations of the two quantities considered. The following simple case considered by Kapteyn may help to make this point clear. Let the variations of x and y be due to a number $m + n$ of elementary causes, m of these causes being common to both x and y, while the remaining n causes are independent. Then we may write

$$x = A_1\alpha_1 + A_2\alpha_2 + \ldots + A_m\alpha_m + B_1\beta_1 + B_2\beta_2 + \ldots + B_n\beta_n,$$

$$y = D_1\alpha_1 + D_2\alpha_2 + \ldots + D_m\alpha_m + C_1\gamma_1 + C_2\gamma_2 + \ldots + C_n\gamma_n,$$

where $\alpha_1, \alpha_2, \ldots, \beta_1, \beta_2, \ldots, \gamma_1, \gamma_2, \ldots$, are all independent of one another. We shall further simplify the problem by supposing that the M.S.E. introduced into x or y by any one of the independent elementary causes is unity. Then

$$x = \overset{m}{\underset{1}{\Sigma}} \alpha_s + \overset{n}{\underset{1}{\Sigma}} \beta_s,$$

$$y = \overset{m}{\underset{1}{\Sigma}} \alpha_s + \overset{n}{\underset{1}{\Sigma}} \gamma_s,$$

where the M.S.E. of each variable α, β, or γ is unity. Then if ϵ_x, ϵ_y be the M.S.E. of x and y respectively

$$\epsilon_x{}^2 = m + n, \quad \epsilon_y{}^2 = m + n,$$

$$xy = \overset{m}{\underset{1}{\Sigma}} \alpha_s{}^2 + \text{products of the form } \alpha_s \alpha_t,\ \alpha\beta,\ \alpha\gamma,\ \beta\gamma.$$

When mean values are taken all the products vanish

$$\overline{xy} = m,$$

$$r \epsilon_x \epsilon_y = r(m + n) = n.$$

Hence $r = \dfrac{m}{m + n}$, and in this particular case r measures the proportion of elementary causes of variation which the two variables have in common.

79. Number of Significant Figures to be used in Computation.

For most practical purposes it is sufficient to work in units which give a range of about 25 in the observations, say from -12 to $+12$ about the mean. In the computation on page 159 the observations were available to the number of significant figures shown in the first two columns. It is now suggested that the data for duration of eclipse could have been rounded off to the nearest whole number, and the orbital period could have been rounded off to the nearest ten, and written down in units of ten days. The orbital periods then range from 1 to 25 units, and the duration of eclipse from 3 to 26 units. The mean orbital period is 7·9 units, and the y's are taken to be deviations from 8 units. The value of Σy^2 is 1146, or corrected for true mean, $1146 - 38 \times \cdot 09$ or 1143. The mean duration comes out to 10·6, and the x's are computed about 10. The value of Σx^2 is 1068, or corrected for true mean, $1086 - 38 \times \cdot 36$ or 1074. The product terms add up to 967, or corrected for true means, $967 - 38 \times \cdot3 \times \cdot6$ or 960. Hence the coefficient of correlation is given by

$$r = \frac{960}{\sqrt{1143 \times 1074}} = \cdot87.$$

The error in the correlation coefficient introduced by the rounding off of the figures is thus $\cdot01$, which is negligible by comparison with the probable error

$$\cdot6745 \frac{\sqrt{1 - (\cdot88)^2}}{\sqrt{38}} \text{ or } \cdot09.$$

If the observations vary through a range of say $- 50$ to $+ 50$ units about the mean, or through a complete range of 100 units, the individual observations may be divided by 4, rounding off the quotients to the nearest unit. The effect of rounding off on the magnitude of r will usually be very small, and the gain in time will be very considerable, since squares and products can then be written down by inspection, and the whole computation can be carried out in a few minutes. The example computed in the table on page 159 above was originally carried out as shown there, and it has been left in the original form in order that the result might be compared with the results derived by other methods, including the results of rounding off the figures as suggested above.

For a discussion of this and other aspects of the computation and use of correlation coefficients the reader is referred to a paper by Walker and Bliss in the *Quarterly Journal of the Royal Meteorological Society*, Vol. 52, page 73.

80. Partial Correlation of three or more variables.

In the earlier sections of the present chapter we have been concerned with the correlation of two variables. This is insufficient for the discussion of many problems in which more than two correlated variables are involved, and the methods of partial correlation have been developed to deal with such cases. The notation used below is due to Udny Yule [*].

In the general case there are n correlated variables

$$X_1, X_2, X_3, \ldots, X_n.$$

The suffixes are here used to distinguish the *variables*, and not the different *observations* of the same variable. This involves a change

---

[*] Udny Yule, *Theory of Statistics.*

from our earlier notation, but as this notation is in general use in discussions of partial correlation, and is the only convenient notation for dealing with the most general cases, it is adopted here in spite of its variation from the practice of the earlier part of this book.

The problem has a two-fold aspect. In the first place we require a method of finding a regression equation to connect one of the variables, say X_1, with all the other variables, in such a form as

$$X_1 = a + b_2 X_2 + b_3 X_3 + \ldots + b_n X_n.$$

The existence of an appreciable coefficient b_s in this equation will denote a correlation between X_1 and X_s which is positive or negative according as b_s is positive or negative. In the second place we require to find the correlation between X_1 and each variable X_s when all the other variables have been allowed for. In practice we work not with the original variables X_1, X_2, etc., but with the deviations x_1, x_2, etc. of these variables from their mean values. Then the regression equation may be written

$$x_1 = b_{12} x_2 + b_{13} x_3 + \ldots + b_{1n} x_n,$$

where each coefficient has two suffixes, the first being the same as that of the variable on the left-hand side of the equation, and the second indicating the variable to which the coefficient is attached. The constant term a must vanish, since the mean value of each term in the above equation is zero. Note that the regression equation for x_2 will be

$$x_2 = b_{21} x_1 + b_{23} x_3 + \ldots + b_{2n} x_n,$$

where b_{21} is not in general the same as b_{12}. The adjustment of the constants may be carried out by the method of least squares, but it becomes tedious when the number of variables, n, is great. We shall only deal in detail with the cases of three and four variables.

Three Variables.

Let the three variables be x_1, x_2, and x_3, and let r_{12} be the correlation coefficient between the variables x_1 and x_2 as calculated directly from the observations, r_{23} and r_{31} being similar correlation coefficients between x_2 and x_3, and between x_3 and x_1. Let $\sigma_1, \sigma_2, \sigma_3$ be the standard deviations of x_1, x_2, and x_3.

The regression equation of x_1 on x_2 and x_3 is

$$x_1 = b_{12}x_2 + b_{13}x_3.$$

The condition that

$$\Sigma\,(x_1 - b_{12}x_2 - b_{13}x_3)^2$$

shall be a minimum reduces to

$$\Sigma\,x_2\,(x_1 - b_{12}x_2 - b_{13}x_3) = 0,$$

$$\Sigma\,x_3\,(x_1 - b_{12}x_2 - b_{13}x_3) = 0.$$

These equations may be written

$$r_{12}\sigma_1\sigma_2 - b_{12}\sigma_2{}^2 - b_{13}r_{23}\sigma_2\sigma_3 = 0,$$

$$r_{13}\sigma_1\sigma_3 - b_{12}r_{23}\sigma_2\sigma_3 - b_{13}\sigma_3{}^2 = 0,$$

or

$$r_{12}\sigma_1 - b_{12}\sigma_2 - b_{13}\,r_{23}\sigma_3 = 0,$$

$$r_{13}\sigma_1 - b_{12}r_{23}\sigma_2 - b_{13}\sigma_3 = 0.$$

From these two equations we find, by solving in the usual manner, that

$$b_{12} = \frac{\sigma_1}{\sigma_2}\frac{r_{12} - r_{13}r_{23}}{1 - r^2{}_{23}},$$

$$b_{13} = \frac{\sigma_1}{\sigma_3}\frac{r_{13} - r_{12}r_{23}}{1 - r^2{}_{23}}.$$

Substitution of these values in the regression equation gives the final form of that equation.

We further require to know the correlation between x_1 and x_2 when each has been corrected for the correlation with x_3. This is written $r_{12\cdot3}$, indicating that the effect of the variable x_3 is eliminated. Since the regression equations of x_1 and x_2 on x_3 are

$$x_1 - r_{13}\frac{\sigma_1}{\sigma_3}x_3 = 0,$$

$$x_2 - r_{23}\frac{\sigma_2}{\sigma_3}x_3 = 0,$$

$r_{12\cdot3}$ is the coefficient of correlation between

$$x_1 - r_{13}\frac{\sigma_1}{\sigma_3}x_3 \quad \text{and} \quad x_2 - r_{23}\frac{\sigma_2}{\sigma_3}x_3.$$

The standard deviations of these two quantities are known (see page 156 last line) to be $\sigma_1(1-r^2_{13})^{\frac{1}{2}}$ and $\sigma_2(1-r^2_{23})^{\frac{1}{2}}$. Hence

$$r_{12 \cdot 3}\,\sigma_1\sigma_2\,(1-r^2_{13})^{\frac{1}{2}}(1-r^2_{23})^{\frac{1}{2}}$$

= mean value of

$$\left(x_1 - r_{13}\frac{\sigma_1}{\sigma_3}\,x_3\right)\left(x_2 - r_{23}\frac{\sigma_2}{\sigma_3}\,x_3\right)$$

= mean value of

$$x_1 x_2 - r_{23}\frac{\sigma_2}{\sigma_3}\,x_1 x_3 - r_{13}\frac{\sigma_1}{\sigma_3}\,x_2 x_3 + r_{13}r_{23}\frac{\sigma_1\sigma_2}{\sigma_3^2}\,x_3^2$$

$$= r_{12}\sigma_1\sigma_2 - r_{13}r_{23}\sigma_1\sigma_2 - r_{13}r_{23}\sigma_1\sigma_2 + r_{13}r_{23}\sigma_1\sigma_2$$

$$= \sigma_1\sigma_2\,(r_{12} - r_{13}r_{23}).$$

Hence
$$r_{12 \cdot 3} = \frac{r_{12} - r_{13}r_{23}}{(1-r^2_{13})^{\frac{1}{2}}(1-r^2_{23})^{\frac{1}{2}}}.$$

Similar forms may be written down for $r_{23 \cdot 1}$ and $r_{13 \cdot 2}$. The line of regression of $x_1 - r_{13}\dfrac{\sigma_1}{\sigma_3}\,x_3$ on $x_2 - r_{23}\dfrac{\sigma_2}{\sigma_3}\,x_3$ is

$$x_1 - r_{13}\frac{\sigma_1}{\sigma_3}x_3 - r_{12 \cdot 3}\frac{\sigma_1(1-r^2_{13})^{\frac{1}{2}}}{\sigma_2(1-r^2_{23})^{\frac{1}{2}}}\left(x_2 - r_{23}\frac{\sigma_2}{\sigma_3}x_3\right) = 0,$$

and it can readily be shown, by substitution for $r_{12 \cdot 3}$, that this is identical, as of course it must be, with the line

$$x_1 - b_{12}x_2 - b_{13}x_3 = 0.$$

Using the equation of p. 156 above, we therefore find the standard deviation of the estimate of x_1 derived from the regression equation is

$$\sigma_1(1-r^2_{13})^{\frac{1}{2}}(1-r^2_{12 \cdot 3})^{\frac{1}{2}}$$

usually denoted $\sigma_{1 \cdot 23}$. This is also equal to

$$\sigma_1(1-r^2_{12})^{\frac{1}{2}}(1-r^2_{13 \cdot 2})^{\frac{1}{2}},$$

since we may take x_2 or x_3 first.

If R is the correlation between the two sides of the regression equation, then

$$\sigma^2_{1 \cdot 23} = \sigma_1^2\,(1 - R^2).$$

Hence
$$1 - R^2 = (1-r^2_{13})(1-r^2_{12 \cdot 3}) = (1-r^2_{12})(1-r^2_{13 \cdot 2}).$$

Four Variables.

The formulae derived above may be extended to four or more variables. If the variables are x_1, x_2, x_3, x_4, we can apply the same methods to eliminate the effects of x_3 and x_4, and find the correlation between x_1 and x_2 which then results. The coefficient of correlation between x_1 and x_2 is then written

$$r_{12\cdot34} = \frac{r_{12\cdot4} - r_{13\cdot4}\, r_{23\cdot4}}{(1 - r^2_{13\cdot4})^{\frac{1}{2}}\,(1 - r^2_{23\cdot4})^{\frac{1}{2}}},$$

the coefficients $r_{12\cdot4}$, $r_{13\cdot4}$, $r_{23\cdot4}$ being derived by the standard method used for three variables above.

The last equation can also be reduced to the form

$$r_{12\cdot34} = \frac{r_{12}\,(1 - r^2_{34}) - r_{13}\,(r_{23} - r_{24}r_{34}) - r_{14}\,(r_{24} - r_{23}r_{34})}{(1 - r^2_{13} - r^2_{14} - r^2_{34} + 2r_{13}r_{14}r_{34})^{\frac{1}{2}}\,(1 - r^2_{23} - r^2_{24} - r^2_{34} + 2r_{23}r_{24}r_{34})^{\frac{1}{2}}},$$

which is a convenient form for computation.

The regression equation for x_1 is

$$x_1 = b_{12\cdot34}\, x_2 + b_{13\cdot24}\, x_3 + b_{14\cdot34}\, x_4.$$

In evaluating $r_{12\cdot34}$ we virtually form the regression equations of x_1 on x_3, x_4 and of x_2 on x_3, x_4,

$$x_1 = b_{13\cdot4}\, x_3 + b_{14\cdot3}\, x_4,$$
$$x_2 = b_{23\cdot4}\, x_3 + b_{24\cdot3}\, x_4.$$

The standard deviations of

$$x_1 - b_{13\cdot4}\, x_3 - b_{14\cdot3}\, x_4 \quad \text{and} \quad x_2 - b_{23\cdot4}\, x_3 - b_{24\cdot3}\, x_4$$

are, as shown in the discussion of three variables,

$$\sigma_1\,(1 - r^2_{14})^{\frac{1}{2}}\,(1 - r^2_{13\cdot4})^{\frac{1}{2}}$$
and
$$\sigma_2\,(1 - r^2_{24})^{\frac{1}{2}}\,(1 - r^2_{14\cdot3})^{\frac{1}{2}}.$$

The regression equation of the first of these expressions on the other will therefore yield a standard deviation

$$\sigma_{1\cdot234} = \sigma_1\,(1 - r^2_{14})^{\frac{1}{2}}\,(1 - r^2_{13\cdot4})^{\frac{1}{2}}\,(1 - r^2_{12\cdot34})^{\frac{1}{2}}$$
$$= \sigma_1\,(1 - r^2_{12})^{\frac{1}{2}}\,(1 - r^2_{13\cdot2})^{\frac{1}{2}}\,(1 - r^2_{14\cdot23})^{\frac{1}{2}}$$
$$= \sigma_1\,(1 - r^2_{13})^{\frac{1}{2}}\,(1 - r^2_{14\cdot3})^{\frac{1}{2}}\,(1 - r^2_{12\cdot34})^{\frac{1}{2}},$$

the regression equation itself being

$$x_1 - b_{13 \cdot 4} \, x_3 - b_{14 \cdot 3} \, x_4$$

$$- r_{12 \cdot 34} \frac{\sigma_1 (1 - r^2_{14})^{\frac{1}{2}} (1 - r^2_{13 \cdot 4})^{\frac{1}{2}}}{\sigma_2 (1 - r^2_{24})^{\frac{1}{2}} (1 - r^2_{23 \cdot 4})^{\frac{1}{2}}} \, (x_2 - b_{23 \cdot 4} \, x_3 - b_{24 \cdot 3} x_4) = 0.$$

Hence

$$b_{12 \cdot 34} = r_{12 \cdot 34} \frac{\sigma_1}{\sigma_2} \left(\frac{(1 - r^2_{14})(1 - r^2_{13 \cdot 4})}{(1 - r^2_{24})(1 - r^2_{23 \cdot 4})} \right)^{\frac{1}{2}}.$$

The other coefficients may be expressed in a variety of forms, but perhaps the most convenient is that given here for $b_{12 \cdot 34}$. The coefficients $b_{13 \cdot 24}$, $b_{14 \cdot 23}$ are obtained by interchanging the suffixes 2 and 3 in the one case and 2 and 4 in the other.

Similar methods can be developed for larger numbers of variables, but the labour involved is very great when more than four variables are involved. An approximate method for use in such cases has been developed by Brooks*.

* Meteorological Office Professional Notes, No. 47.

CHAPTER XI

HARMONIC ANALYSIS

81. Fourier's Theorem.

The general problem of the representation of an arbitrary function of a variable θ, say $f(\theta)$, by a trigonometrical series of the form

$$\left.\begin{aligned}f(\theta) = A_0 + A_1 \cos \theta + A_2 \cos 2\theta + A_3 \cos 3\theta + \dots \\ + B_1 \sin \theta + B_2 \sin 2\theta + B_3 \sin 3\theta + \dots\end{aligned}\right\}\dots(1),$$

was first discussed by Fourier in his work on the conduction of heat. Fourier showed that in a large number of particular cases the trigonometrical series actually converged to the sum $f(\theta)$, but the first rigorous general proof was given by Dirichlet. The coefficients A_0, A_1, A_2, B_1, B_2, etc. are defined as follows. Integrating both sides of equation (1) between the limits 0 and 2π, we obtain

$$\int_0^{2\pi} f(\theta)\, d\theta = 2\pi A_0 \quad \dots\dots\dots\dots\dots\dots(2).$$

Again multiplying both sides of equation (1) by $\cos s\theta$ and integrating between the limits 0 and 2π, we find

$$\int_0^{2\pi} f(\theta) \cos s\theta\, d\theta = A_s \int_0^{2\pi} \cos^2 s\theta\, d\theta = \pi A_s \quad \dots\dots(3),$$

since
$$\int_0^{2\pi} \cos s\theta \cdot \cos r\theta\, d\theta = 0, \quad \int_0^{2\pi} \cos s\theta \cdot \sin r\theta\, d\theta = 0.$$

Similarly

$$\int_0^{2\pi} f(\theta) \sin s\theta\, d\theta = \pi B_s \quad \dots\dots\dots\dots\dots\dots(4).$$

Equations (2), (3), and (4) determine the values of the coefficients in equation (1). These four equations constitute a definition of Fourier's Theorem. For a full account of the analytical conditions which $f(\theta)$ must satisfy in order that equation (1) may be true, the reader is referred to Whittaker and Watson's *Modern Analysis*, Chapter IX.

82. Fourier Series with a Limited Number of Observations.

In practice $f(\theta)$ is not given for all values of θ from 0 to 2π, but is given for a series of equidistant values of θ. For example θ may represent time, and we may have a series of hourly observations which we require to represent by a trigonometrical series, each term of which has a period which is either 24 hours or a sub-multiple of 24 hours.

In the general case let there be n observations $l_0, l_1, l_2, ..., l_{n-1}$, associated with equidistant scale values of a variable θ; l_r corresponds to $\theta_r = r\dfrac{2\pi}{n}$. The problem is to find the best representation of the observations by a series of the form

$$\left.\begin{aligned}l_r = A_0 + A_1\cos\theta_r + A_2\cos 2\theta_r + ... + A_m\cos m\theta_r \\ + B_1\sin\theta_r + B_2\sin 2\theta_r + ... + B_m\sin m\theta_r\end{aligned}\right\}\ \ ......(5),$$

or $\qquad l_r = A_0 + \sum_{s=1}^{s=m} A_s\cos s\theta_r + \sum_{i=1}^{i=m} B_s\sin s\theta_r$

$$= A_0 + \sum_{s=1}^{s=m} R_s\cos(s\theta_r - \phi_s)\ ....(5)'.$$

If the Fourier series represented in equation (5) extends to n terms, then the number of unknown coefficients is equal to the number of observations, and a complete representation of the n observations by the series is possible. In any case $2r + 1$ cannot exceed n.

We now have to replace equations (2), (3) and (4) by equations in which integration is replaced by summation. If we put $d\theta = \dfrac{2\pi}{n}$, and replace $f(\theta)$ by l_r, and θ by $\theta_r = r\dfrac{2\pi}{n}$, equations (2), (3) and (4) become

$$A_0 = \frac{1}{n}\sum_{r=0}^{r=n-1} l_r\ ....(6),$$

$$A_s = \frac{2}{n}\sum_{r=0}^{r=n-1} l_r\cos r\theta_s\ ....(7),$$

$$B_s = \frac{2}{n}\sum_{r=0}^{r=n-1} l_r\sin r\theta_s\ ....(8).$$

Equations (7) and (8) depend upon the assumption that

$$\sum_{r=0}^{r=n-1} \sin^2 r\theta_s = \sum_{r=0}^{r=n-1} \cos^2 r\theta_s = \frac{n}{2} \quad \dots\dots\dots\dots(9).$$

This is true in all possible cases* except when $r\theta_s$ is zero or a multiple of π for all values of r; in such cases $\cos^2 r\theta_s = 1$, and the right-hand side of equation (7) has the factor $\frac{1}{n}$ instead of $\frac{2}{n}$ The case of $r\theta_s = 0$ corresponds to the computation of A_0, for which we have already derived the factor $\frac{1}{n}$. The second possibility corresponds to $\theta_s = \pi$ or $s = \frac{n}{2}$. When n is an odd number this contingency cannot arise, but when n is an even number it follows that $B_{\frac{n}{2}}$ vanishes, and

$$A_{\frac{n}{2}} = \frac{1}{n} \sum_{r=0}^{r=n-1} (-1)^r l_r \quad \dots\dots\dots\dots(10).$$

When n is odd, all the coefficients are derivable by the straightforward use of equations (7) and (8). The coefficients of the trigonometrical terms in equation (5) are therefore all determined by equations (6), (7), and (8), except for the final coefficient in a series for which n is an even number, which must be determined by the use of equation (10).

---

* The following series are given in any standard text-book on trigonometry:

$$\cos \alpha + \cos(\alpha + \beta) + \cos(\alpha + 2\beta) + \dots + \cos(\alpha + \overline{n-1}\,\beta) = \cos(\alpha + \tfrac{1}{2}(n-1)\beta)\,\frac{\sin \frac{1}{2}n\beta}{\sin \frac{1}{2}\beta},$$

$$\sin \alpha + \sin(\alpha + \beta) + \sin(\alpha + 2\beta) + \dots + \sin(\alpha + \overline{n-1}\,\beta) = \sin(\alpha + \tfrac{1}{2}(n-1)\beta)\,\frac{\sin \frac{1}{2}n\beta}{\sin \frac{1}{2}\beta}.$$

It follows that

$$\sum_{r=0}^{r=n-1} \sin^2 r\theta_s = \frac{n}{2} - \sum_{r=0}^{r=n-1} \cos 2r\theta_s,$$

$$\sum_{r=0}^{r=n-1} \cos^2 r\theta_s = \frac{n}{2} + \sum_{r=0}^{r=n-1} \cos 2r\theta_s.$$

For the series on the right-hand side of these two equations, $\alpha = 0$; $\beta = 2\theta_s = 4\pi s/n$; $\frac{1}{2}n\beta = 2\pi s$; $\frac{1}{2}\beta = 2\pi s/n$.

Thus $\sin \frac{1}{2}n\beta$ will vanish for all values of s, while $\sin \frac{1}{2}\beta$ is finite except when $s = \frac{1}{2}n$, and the sum of the series will always vanish except when $s = \frac{1}{2}n$, in which case the expression on the right-hand side $\sin \frac{1}{2}n\beta/\sin \frac{1}{2}\beta$ is indeterminate. But $\sin^2 r\theta_s$ is then zero, and $\cos^2 r\theta_s$ unity for all values of r, so that $\Sigma \sin^2 r\theta_s = 0$, and $\Sigma \cos^2 r\theta_s = n$.

83. Harmonic Analysis from the Standpoint of Least Squares.

We might also start from equation (5) and obtain the values of the coefficients by the method of least squares. The method yields the same equations (6), (7), (8) and (10) as were derived above. The values of the $(2m + 1)$ coefficients in equation (5) have to be chosen so as to make

$$\sum_{r=0}^{r=n-1} \left(l_r - A_0 - \sum_{s=1}^{s=m} A_s \cos s\theta_r - \sum_{s=1}^{s=m} B_s \sin s\theta_r \right)^2 \quad \dots(11)$$

a minimum. Differentiating this expression with respect to each of the coefficients in turn, we obtain the normal equations

$$\sum_{r=0}^{r=n-1} \left(l_r - A_0 - \sum_{s=1}^{s=m} A_s \cos s\theta_r - \sum_{s=1}^{s=m} B_s \sin s\theta_r \right) = 0 \dots(12)$$

with m pairs of relations of the form

$$\sum_{r=0}^{r=n-1} \cos s\theta_r \left(l_r - A_0 - \sum_{s=1}^{s=m} A_s \cos s\theta_r - \sum_{s=1}^{s=m} B_s \sin s\theta_r \right) = 0 \dots(13),$$

$$\sum_{r=0}^{r=n-1} \sin s\theta_r \left(l_r - A_0 - \sum_{s=1}^{s=m} A_s \cos s\theta_r - \sum_{s=1}^{s=m} B_s \sin s\theta_r \right) = 0 \dots(14).$$

In these equations $\theta_r = 2\pi r/n$. In equation (12) A_s and B_s are multiplied by $\sum_{r=0}^{r=n-1} \cos 2\pi rs/n$ and $\sum_{r=0}^{r=n-1} \sin 2\pi rs/n$ respectively. It follows (see footnote, page 181) that these series vanish, and equation (12) reduces to

$$\sum_{r=0}^{r=n-1} l_r - nA_0 = 0 \quad \dots\dots\dots\dots\dots(6)',$$

which is equation (6) already derived above.

In equation (13) A_0 is multiplied by $\Sigma \cos s\theta_r$ or $\Sigma \cos 2\pi rs/n$, which is zero; A_s is multiplied by $\sum_{r=0}^{r=n-1} \cos^2 2\pi rs/n$, which we have found to be equal to $\frac{1}{2}n$; $A_p (p \neq s)$ is multiplied by

$$\sum_{r=0}^{r=n-1} \cos 2\pi rs/n \cos 2\pi pr/n$$

or　　　$\frac{1}{2} \sum_{r=0}^{r=n-1} \{\cos (p - s) 2\pi r/n + \cos (p + s) 2\pi r/n\},$

and both these series vanish as is seen by comparing with the series in footnote, page 181; B_p is multiplied by

$$\sum_{r=0}^{r=n-1} \cos 2\pi \, rs/n \, \sin 2\pi \, pr/n,$$

or $\qquad \frac{1}{2} \sum_{r=0}^{r=n-1} \{\sin(p+s)2\pi \, r/n - \sin(p-s)2\pi \, r/n\},$

and both the series vanish for all values of p, including the case of $p = s$. Thus equation (13) reduces to

$$\frac{n}{2} A_s = \sum_{r=0}^{r=n-1} l_r \cos s\theta_r \qquad \ldots\ldots\ldots\ldots\ldots(7)'.$$

In the same way it is readily shown that in equation (14) the only terms which remain are those in l and B_s, giving

$$\frac{n}{2} B_s = \sum_{r=0}^{r=n-1} l_r \sin s\theta_r \qquad \ldots\ldots\ldots\ldots\ldots(8)'.$$

When the number of observations is even, and the series is carried on to $A_{\frac{n}{2}}$, then in the normal equation for $A_{\frac{n}{2}}$ we shall find $A_{\frac{n}{2}}$ multiplied by $\sum_{r=0}^{r=n-1} \cos^2 \dfrac{nr}{2}\dfrac{2\pi}{n}$ or $\sum_{r=0}^{r=n-1} \cos^2 r\pi$, each term of which is unity. Hence $A_{\frac{n}{2}}$ is given by

$$n A_{\frac{n}{2}} = \sum_{r=0}^{r=n-1} (-1)^r l_r \qquad \ldots\ldots\ldots\ldots\ldots(10)'.$$

No term in $B_{\frac{n}{2}}$ can be derived, as $\Sigma \sin^2 r\pi = 0$, and the sine-terms end with $B_{\frac{n}{2}-1}$.

Thus equations (6), (7), (8) and (10) are the normal equations derivable by the methods of least squares, and the straightforward application of these equations will enable us to derive the co-efficients which we may require to evaluate.

The series may also be written in the form

$$l = A_0 + R_1 \sin(\theta - \phi_1) + R_2 \sin(2\theta - \phi_2) + \ldots \ldots (15),$$

where $\qquad \left. \begin{array}{l} R_s^2 = A_s^2 + B_s^2, \quad \tan \phi_s = \dfrac{B_s}{A_s} \\[2mm] R_s \cos \phi_s = A_s, \quad R_s \sin \phi_s = B_s \end{array} \right\} \qquad \ldots\ldots\ldots\ldots(16).$

84. Accuracy of the Representation by a Series with a Finite Number of Terms.

Let the n observations $l_0, l_1, \ldots, l_{n-1}$ have a standard deviation σ, defined by

$n\sigma^2 =$ sum of squares of deviations of the l's from their mean value.

The effect of correcting the l's for a harmonic term $R_s \cos(s\theta - \phi_s)$ is to give a series of quantities u_0, u_1, etc.,

$$u = l - A_0 - R_s \cos(s\theta - \phi_s).$$

The mean value of u is zero, since the mean value of the term in R_s is zero in a complete cycle. Hence if σ is the standard deviation of the u's,

$$n\sigma'^2 = \Sigma u^2 = \Sigma (l - A_0 - R_s \cos(s\theta - \phi_s))^2$$
$$= \Sigma\{l^2 + A_0^2 - 2lA_0 + R_s^2 \cos^2(s\theta - \phi_s) - 2lR_s \cos(s\theta - \phi_s)$$
$$+ 2A_0 R_s \cos(s\theta - \phi_s)\}$$
$$= \Sigma l^2 + nA_0^2 - 2nA_0^2 + \tfrac{1}{2}nR_s^2 - 2R_s\Sigma l \cos(s\theta - \phi_s) + 0$$
$$= \Sigma l^2 - nA_0^2 + \tfrac{1}{2}nR_s^2$$
$$= n(\sigma^2 - \tfrac{1}{2}R_s^2).$$

Hence $\qquad\qquad\qquad \sigma'^2 = \sigma^2 - \tfrac{1}{2}R_s^2 \qquad\qquad$ ....(17).

In deriving the value of $n\sigma'^2$ above use is made of equation (6) to express $-2lA_0$ as $-nA_0^2$; the mean value of $\cos^2(s\theta - \phi_s)$ is $\tfrac{1}{2}$ in a complete cycle, except when n is even and

$$s = \frac{n}{2};$$

$$\Sigma l \cos(s\theta - \phi_s) = \cos\phi_s \Sigma l \cos s\theta + \sin\phi_s \Sigma l \sin s\theta$$
$$= \frac{n}{2}(A_s \cos\phi_s + B_s \sin\phi_s) = \frac{n}{2}R_s$$

from (16). Also

$$\Sigma \cos(s\theta - \phi_s) = 0.$$

In the exceptional case of $s = \dfrac{n}{2}$, the terms

$$R_s^2 \cos^2(s\theta - \phi_s) - 2lR_s \cos(s\theta - \phi_s)$$

reduce to $\qquad\qquad nR_{\frac{n}{2}}^2 - 2nR_{\frac{n}{2}}^2 \quad$ or $\quad -nR_{\frac{n}{2}}^2$

and equation (17) becomes

$$\sigma'^2 = \sigma^2 - R_{\frac{n}{2}}^2 \qquad\qquad$$ ....(18).

Equation (17) may be applied successively to correct for the effect of any number of terms, and the general result may be stated thus. The standard deviation of the numbers obtained after correcting the observations for any number of terms of the form $R_s \cos(s\theta - \phi_s)$ is

$$\sigma'^2 = \sigma^2 - \tfrac{1}{2}\Sigma R_s^2$$
$$= \sigma^2 - \tfrac{1}{2}\Sigma(A_s^2 + B_s^2)\ \dots\dots\dots\dots\dots(19),$$

with the one exception that if a term in R_n is considered, the appropriate term in (19) is $-\dfrac{R_n^2}{2}$. If the series is carried to n terms, the representation of the observations by the series is accurate, and σ' is then zero. It follows that for n odd

$$\sigma^2 = \tfrac{1}{2}\sum_{s=1}^{\frac{n}{2}-1}(A_s^2 + B_s^2) = \tfrac{1}{2}\sum_{s=1}^{\frac{n-1}{2}} R_s^2 \ \dots\dots\ \dots\dots(20),$$

when n is even

$$\sigma^2 = \tfrac{1}{2}\sum_{s=1}^{\frac{n}{2}-1}(A_s^2 + B_s^2) + A_{\frac{n}{2}}^2 \ \dots\dots\dots\ (21).$$

Equation (19) is useful in testing the degree of accuracy with which the observations are represented by a small number of terms, and in cases when the series is carried on to the full number of terms, the use of equation (20) or (21) gives a useful test of the accuracy of the arithmetic

Just as equations (6), (7), (8), and (10) represent the equivalent for a finite number of observations of Fourier's theorem relating to a function given for all values of the variable, so equations (20) and (21) represent the equivalent of the Hurwitz-Liapounoff theorem* which states that, provided $f(x)$ is bounded and satisfies certain conditions laid down by Dirichlet, then

$$\tfrac{1}{2}A_0^2 + \sum_{r=1}^{\infty}(A_r^2 + B_r^2) \text{ converges to the sum } \frac{1}{\pi}\int_0^{2\pi} \{f(\theta)\}^2\, d\theta,$$

where the coefficients A_0, A_1, B_1, etc. are defined by equations (1) to (4) above.

* See Whittaker and Watson, *Modern Analysis*, Chapter ix.

85. The Evaluation of the Coefficients.

(a) n an odd number.

The coefficients A_0, A_1, A_2, etc. are to be computed by means of equations of the form

$$n A_0 = l_0 + l_1 + l_2 + \ldots\ldots\ldots\ldots\ldots\ldots + l_{n-1},$$

$$\tfrac{1}{2}n A_1 = l_0 + l_1 \cos \frac{2\pi}{n} + l_2 \cos 2 . \frac{2\pi}{n} + \ldots + l_{n-1} \cos (n-1) \frac{2\pi}{n},$$

$$\tfrac{1}{2}n A_2 = l_0 + l_1 \cos 2 . \frac{2\pi}{n} + l_2 \cos 4 . \frac{2\pi}{n} + \ldots + l_{n-1} \cos 2(n-1) \frac{2\pi}{n},$$

$$\tfrac{1}{2}n B_1 = \quad l_1 \sin \frac{2\pi}{n} + l_2 \sin 2 . \frac{2\pi}{n} + \ldots + l_{n-1} \sin (n-1) \frac{2\pi}{n},$$

$$\tfrac{1}{2}n B_2 = \quad l_1 \sin 2 . \frac{2\pi}{n} + l_2 \sin 4 . \frac{2\pi}{n} + \ldots + l_{n-1} \sin 2(n-1) \frac{2\pi}{n},$$

etc.

In these expressions the argument of the cosine or sine which multiplies l_{n-1} differs from a multiple of 2π by the amount of the argument of the cosine or sine which multiplies l_1. Hence the following procedure is adopted. Write the l's thus:

$$l_0 \quad l_1 \quad l_2 \quad \ldots \quad l_{\frac{n-1}{2}},$$

$$l_{n-1} \quad l_{n-2} \quad \ldots \quad l_{\frac{n+1}{2}}.$$

Add the rows $a_0 \quad a_1 \quad a_2 \quad \ldots \quad a_{n-1}.$

Subtract the rows $b_1 \quad b_2 \quad \ldots \quad b_{n-1}.$

The coefficients may then be derived from the following equations

$$n A_0 = a_0 + a_1 \qquad + a_2 \qquad + \ldots + a_{\frac{n-1}{2}},$$

$$\tfrac{1}{2}n A_1 = a_0 + a_1 \cos \frac{2\pi}{n} + a_2 \cos 2 . \frac{2\pi}{n} + \ldots + a_{\frac{n-1}{2}} \cos \frac{n-1}{2} . \frac{2\pi}{n},$$

$$\tfrac{1}{2}n A_2 = a_0 + a_1 \cos 2 . \frac{2\pi}{n} + a_2 \cos 4 . \frac{2\pi}{n} + \ldots + a_{\frac{n-1}{2}} \cos 2 . \frac{n-1}{2} . \frac{2\pi}{n},$$

$$\tfrac{1}{2}n B_1 = \quad b_1 \sin \frac{2\pi}{n} + b_2 \sin 2 . \frac{2\pi}{n} + \ldots + b_{\frac{n-1}{2}} \sin \frac{n-1}{2} . \frac{2\pi}{n},$$

$$\tfrac{1}{2}n B_2 = \quad b_1 \sin 2 . \frac{2\pi}{n} + b_2 \sin 4 . \frac{2\pi}{n} + \ldots + b_{\frac{n-1}{2}} \sin 2 . \frac{n-1}{2} . \frac{2\pi}{n}.$$

This procedure halves the amount of multiplication.

Turner's tables for use in harmonic analysis give for different values of n and for different values of l the products $l \frac{\cos}{\sin} r \frac{2\pi}{n}$. It is recommended that the quantities a, b should first be formed, and the tables then used for the first half of the entries only.

(b) n an even number ($n = 2p$, p odd).

When n is even and equal to $2p$, l_p will always be multiplied by the cosine or sine of a multiple of π, and the angles whose functions are required are symmetrical about 0 and π, and about $\frac{\pi}{2}$ and $\frac{3\pi}{2}$. The procedure adopted for n odd is slightly modified, and applied twice over. The observations are written thus:

$$l_0 \quad l_1 \quad l_2 \quad \ldots \quad l_{p-1} \quad l_p,$$
$$l_{2p-1} \quad l_{2p-2} \quad \ldots \quad l_{p+1}.$$

Add the rows $a_0 \quad a_1 \quad a_2 \quad \ldots \quad a_{p-1} \quad a_p.$

Subtract the rows $b_1 \quad b_2 \quad \ldots \quad b_{p-1}.$

Again write the a's in the form

$$a_0 \quad a_1 \quad a_2 \quad \ldots \quad a_{\frac{p-1}{2}},$$
$$a_p \quad a_{p-1} \quad a_{p-2} \quad \ldots \quad a_{\frac{p+1}{2}}.$$

Add the rows $a_0 \quad \alpha_1 \quad \alpha_2 \quad \ldots \quad \alpha_{\frac{p-1}{2}}.$

Subtract the rows $\alpha_0' \quad \alpha_1' \quad \alpha_2' \quad \ldots \quad \alpha'_{\frac{p-1}{2}}.$

Again write the b's in the form

$$b_1 \quad b_2 \quad \ldots \quad b_{\frac{p-1}{2}},$$
$$b_{p-1} \quad b_{p-2} \quad \ldots \quad b_{\frac{p+1}{2}}.$$

Add the rows $\beta_1 \quad \beta_2 \quad \ldots \quad \beta_{\frac{p-1}{2}}.$

Subtract the rows $\beta_1' \quad \beta_2' \quad \ldots \quad \beta'_{\frac{p-1}{2}}.$

The coefficients can then be readily deduced from the equations

$$4nA_0 = \alpha_0 + \alpha_1 + \alpha_2 + \text{etc.},$$

$$2nA_1 = \alpha_0' + \alpha_1' \cos\frac{2\pi}{n} + \alpha_2' \cos 2.\frac{2\pi}{n} + \ldots,$$

$$2nA_2 = \alpha_0 + \alpha_1 \cos 2.\frac{2\pi}{n} + \alpha_2 \cos 4.\frac{2\pi}{n} + \ldots,$$

$$2nA_3 = \alpha_0' + \alpha_1' \cos 3.\frac{2\pi}{n} + \alpha_2' \cos 6.\frac{2\pi}{n} + \ldots,$$

$$2nA_4 = \alpha_0 + \alpha_1 \cos 4.\frac{2\pi}{n} + \alpha_2 \cos 8.\frac{2\pi}{n} + \ldots,$$

$$2nB_1 = \beta_1 \sin\frac{2\pi}{n} + \beta_2 \sin 2.\frac{2\pi}{n} + \beta_3 \sin 3.\frac{2\pi}{n} + \ldots,$$

$$2nB_2 = \beta_1' \sin 2.\frac{2\pi}{n} + \beta_2' \sin 4.\frac{2\pi}{n} + \beta_3' \sin 6.\frac{2\pi}{n} + \ldots,$$

$$2nB_3 = \beta_1 \sin 3.\frac{2\pi}{n} + \beta_2 \sin 6.\frac{2\pi}{n} + \beta_3 \sin 9.\frac{2\pi}{n} + \ldots,$$

$$2nB_4 = \beta_1' \sin 4.\frac{2\pi}{n} + \beta_2' \sin 8.\frac{2\pi}{n} + \beta_-\quad 12.\frac{2\pi}{n} + \ldots.$$

Note that the accented and unaccented α's and β's are used alternately in finding successive A's and B's.

The number of multiplications involved in finding any one coefficient is never greater than $\frac{1}{4}n + 1$.

If only the first harmonic is required, only $\alpha_0', \alpha_1', \ldots, \beta_1, \beta_2, \ldots$ are evaluated.

(c) n a multiple of 4 ($n = 4q$).

The procedure is again varied slightly from that used above. Write the observations (the l's) as follows:

	l_0	l_1	l_2	$\ldots$	l_{2q-1}	l_{2q},
		l_{4q-1}			l_{2q+1}.	
Add the rows	a_0	a_1	a_2	$\ldots$	a_{2q-1}	a_{2q}.
Subtract the rows		b_1	b_2	$\ldots$	b_{2q-1}.	

Again write the a's in the form

$$a_0 \quad a_1 \quad a_2 \quad \ldots \quad a_{q-1} \quad a_q,$$

$$a_{2q} \quad a_{2q-1} \quad a_{2q-2} \ldots \quad a_{q+1}.$$

Add the rows $\quad\quad a_0 \quad a_1 \quad a_2 \quad \ldots \quad a_{q-1} \quad a_q.$

Subtract the rows $\quad a_0' \quad a_1' \quad a_2' \quad \ldots \quad a'_{q-1}.$

Similarly write the b's in the form

$$b_1 \quad\quad b_2 \quad\quad \ldots \quad b_{q-1} \quad b_q,$$

$$b_{2q} \quad\quad b_{2q-2} \quad \ldots \quad b_{q+1}.$$

Add the rows $\quad\quad \beta_1 \quad\quad \beta_2 \quad\quad \ldots \quad \beta_{q-1} \quad \beta_q.$

Subtract the rows $\quad \beta_1' \quad\quad \beta_2' \quad\quad \ldots \quad \beta'_{q-1}.$

The coefficients A, B are then derived from the following equations

$$nA_0 = a_0 + a_1 + a_2 + \ldots,$$

$$\tfrac{1}{2}nA_1 = a_0' + a_1' \cos \frac{2\pi}{n} \quad + a_2' \cos 2 . \frac{2\pi}{n} + \ldots,$$

$$\tfrac{1}{2}nA_2 = a_0 \ + a_1 \cos 2 . \frac{2\pi}{n} + a_2 \cos 4 . \frac{2\pi}{n} + \ldots,$$

$$\tfrac{1}{2}nA_3 = a_0' + a_1' \cos 3 . \frac{2\pi}{n} + a_2' \cos 6 . \frac{2\pi}{n} + \ldots,$$

$$\tfrac{1}{2}nA_4 = a_0 \ + a_1 \cos 4 . \frac{2\pi}{n} + a_2 \cos 8 . \frac{2\pi}{n} + \ldots,$$

$$\tfrac{1}{2}nB_1 = \beta_1 \sin \frac{2\pi}{n} \quad + \beta_2 \sin 2 . \frac{2\pi}{n} + \beta_3 \sin 3 . \frac{2\pi}{n} + \ldots,$$

$$\tfrac{1}{2}nB_2 = \beta_1' \sin 2 . \frac{2\pi}{n} + \beta_2' \sin 4 . \frac{2\pi}{n} + \beta_3' \sin 6 . \frac{2\pi}{n} + \ldots,$$

$$\tfrac{1}{2}nB_3 = \beta_1 \sin 3 . \frac{2\pi}{n} + \beta_2 \sin 6 . \frac{2\pi}{n} + \beta_3 \sin 9 . \frac{2\pi}{n} + \ldots,$$

$$\tfrac{1}{2}nB_4 = \beta_1' \sin 4 . \frac{2\pi}{n} + \beta_2' \sin 8 . \frac{2\pi}{n} + \beta_3' \sin 12 . \frac{2\pi}{n} + \ldots,$$

etc.

These equations are in fact identical with those in (b) above, but the a's and β's are not quite identically defined.

(*d*) n a multiple of 8.

Case (*c*) may be slightly extended when n is a multiple of 8, i.e. when q is an even number. The work is carried out as in (*c*) up to the evaluation of the α's and β's. We now write the α's in the form

$$\alpha_0 \quad \alpha_1 \quad \alpha_2 \quad \cdots \qquad \alpha_{\frac{q}{2}},$$

$$\alpha_q \quad \alpha_{q-1} \qquad \cdots \quad \alpha_{\frac{q}{2}+1}.$$

Add the rows $\quad\quad \gamma_0 \quad \gamma_1 \quad \gamma_2 \quad \cdots \qquad \gamma_{\frac{q}{2}}.$

Subtract the rows $\quad \gamma_0' \quad \gamma_1' \quad \gamma_2' \quad \cdots \qquad \gamma'_{\frac{q}{2}}.$

Again write the β'''s in the form

$$\beta_1' \quad \beta_2' \quad \cdots \quad \beta'_{\frac{q}{2}-1} \quad \beta'_{\frac{q}{2}},$$

$$\beta'_{q-1} \ \beta'_{q-2} \quad \cdots \quad \beta'_{\frac{q}{2}+1}.$$

Add the rows $\quad\quad \epsilon_1 \quad \epsilon_2 \quad \cdots \quad \epsilon_{\frac{q}{2}-1} \quad \epsilon_{\frac{q}{2}}.$

Subtract the rows $\quad \epsilon_1' \quad \epsilon_2' \quad \cdots \quad \epsilon'_{\frac{q}{2}-1} \quad \epsilon'^{.}_{\frac{q}{2}}.$

Then the formulae for the even harmonics become

$$n A_0 = \gamma_0 + \gamma_1 + \gamma_2 + \cdots,$$

$$\tfrac{1}{2}n A_2 = \gamma_0' + \gamma_1' \cos 2\frac{2\pi}{n} + \gamma_2' \cos 4\frac{2\pi}{n} + \cdots,$$

$$\tfrac{1}{2}n A_4 = \gamma_0 + \gamma_1 \cos 4\frac{2\pi}{n} + \gamma_2 \cos 8\frac{2\pi}{n} + \cdots,$$

$$\tfrac{1}{2}n A_6 = \gamma_0' + \gamma_1' \cos 6\frac{2\pi}{n} + \gamma_2' \cos 12\frac{2\pi}{n} + \cdots,$$

etc.,

$$\tfrac{1}{2}n B_2 = \epsilon_1 \sin 2\frac{2\pi}{n} + \epsilon_2 \sin 4\frac{2\pi}{n} + \cdots,$$

$$\tfrac{1}{2}n B_4 = \epsilon_1' \sin 4\frac{2\pi}{n} + \epsilon_2' \sin 8\frac{2\pi}{n} + \cdots,$$

$$\tfrac{1}{2}n B_6 = \epsilon_1 \sin 6\frac{2\pi}{n} + \epsilon_2 \sin 12\frac{2\pi}{n} + \cdots,$$

etc.

The odd coefficients must be evaluated by the formulae given at the end of (c) using the α' and β terms.

If only the first harmonic is required, the above procedure is omitted and the case treated as in (c).

86. Computation of R and ϕ given A and B.

For all practical purposes, the computation of R and ϕ is most readily carried out by means of a slide rule. With the ordinary Mannheim type of slide rule this is done as follows. First set 45° on the tangent scale opposite the greater of A and B; then set the cursor to the smaller of these two, and read off the angle on the tangent scale. Let this angle be χ. Leave the cursor in the same position, and move the scale until the angle χ on the sine scale is under the cursor. The reading opposite 90° on the sine-scale then gives the value of R.

The basis of this computation is readily understood. Let A be the greater. Then $\tan \chi = \dfrac{B}{A}$, and $R = \dfrac{B}{\sin \chi}$. If B is greater than A, the formulae become $\tan \chi = \dfrac{A}{B}$, $R = \dfrac{A}{\sin \chi}$.

To determine ϕ, we use the relations

$$R \cos \phi = A, \quad R \sin \phi = B.$$

The quadrant in which ϕ lies is determined by the signs of A and B. Then $\phi = 360° \pm \chi$ or $180° \pm \chi$ if A is numerically greater than B, and $\phi = 90° \pm \chi$ or $270° \pm \chi$ if B is numerically greater than A.

To take an example, let $A = -25$, $B = 30$, $\chi = \tan^{-1} \frac{25}{30} = 40°$, $R = 39$. The angle ϕ is in the second quadrant, and nearer to 90° than to 180°, since B is greater than A. Hence $\phi = 90° + \chi = 130°$.

Care must be taken to use the correct scale for tangents. A simple check is obtained from $\tan 45° = 1\cdot0$, $\tan 30° = 0\cdot58$.

87. Number of Significant Figures to be used in Harmonic Analysis.

Only in special circumstances is it necessary to retain more than two significant figures in the computation of harmonic coefficients, and in most work of this kind it is sufficient that the total range of the l's, the original figures, should amount to say 25. If the original figures show a range of variation of 100 units, the work may occasionally be reduced by dividing all the *l*'s by 4, so changing the unit. But care must be taken to transform the derived amplitudes back into the original units (by multiplying by 4 in this case).

Whether or not it is worth while dividing by an appropriate factor will depend in part upon the number of figures to be analysed. If for example we have 12 monthly mean temperatures to analyse as shown in the table on page 195, we might multiply all the figures by 2, and round off the products to the nearest whole number, but the gain in time would be slight, as the trigonometrical ratios involved are such that all the products can be formed mentally. Such methods of abbreviation are valuable when a series of harmonic analyses have to be evaluated, and the computer has the opportunity to become thoroughly familiar with the order in which the different stages of the work are carried out.

88. Some useful Trigonometrical Ratios.

When 12 observations have to be analysed, the angles which occur in the formulae are all multiples of $30°$. The following relations may be usefully borne in mind as an aid to quick computation:

$$\cos \ 0° = \sin 90° = 1,$$

$$\cos 30° = \sin 60° = \cdot 866 = 1 - \tfrac{1}{10} - \tfrac{1}{30} \text{ approximately,}$$

$$\cos 60° = \sin 30° = \tfrac{1}{2}.$$

For 24 observations the following will be the only additional ratios required:

$$\cos 15° = \sin 75° = \cdot 966 = 1 - \tfrac{1}{30} \text{ approximately,}$$

$$\cos 45° = \sin 45° = \cdot 707,$$

$$\cos 75° = \sin 15° = \cdot 26 = \tfrac{1}{4} + \tfrac{1}{100}.$$

It is thus possible in practically all work with 12 or 24 observations to carry out the formation of the products mentally.

89. Practical Evaluation of Coefficients.

Example 1. Find the first harmonic of the series of 13 monthly observed values:

60   112   125   123   183   189   150   136   125   93   116   96   110.

In order to diminish the magnitude of the numbers with which we shall have to deal, 60 is subtracted from each, and the resulting series is folded upon itself as in (a) above:

	0	52	65	63	123	129	90	
		50	36	56	33	65	76	
Sums (a's)	0	102	101	119	156	194	166	
Cosines	1·0	·87	·57	·12	− ·36	− ·75	− ·97	
Products	0	89	58	14	− 56	− 145	− 161	Sum = − 201
Differences (b's)		2	29	17	90	64	14	
Sines		·47	·82	·99	·94	·66	·24	
Products		1	24	17	85	42	3	Sum = 172

The first harmonic only being required, the sum $\Sigma a \cos \theta$ is evaluated as shown before the b's are formed. The evaluation of R and ϕ can best be done by the use of a slide rule.

$$\phi = 140°,$$

$$R = \frac{2}{13} \times \frac{172}{\sin 40°} = 40.$$

The formation of the products can be carried out without writing in the cosines or sines, if a slide rule is used. If, however, a number of computations have to be carried out for the same value of n, it is worth while writing in the cosines and sines as was done above, as the formation of the products can in most cases be done by inspection.

The amplitude of the first harmonic is therefore 40 units, and the time of occurrence of the maximum is $\frac{140°}{360°} \times 13 = 5$ months after the time of the first observation.

Example 2. To find the first two harmonics of the series of 14 observed values:

11   31   9   87   71   91   67   62   9   63   26   58   36   44.

The whole of the computation can be readily carried out as shown in (b) above:

	11	31	9	87	71	91	67	62
		44	36	58	26	63	9	
a's	11	75	45	145	97	154	76	62
(a's repeated)	62	76	154	97				
Sums a	73	151	199	242				
Differences a'	−51	−1	−109	48				

b's		-13	-27	29	45	28	58
(b's repeated)		58	28	45			
Sums β		45	1	74			
Differences β		-71	-55	-16			

$$14A_0 = \Sigma a = 665, \qquad A_0 = 47\cdot 5,$$

$$7A_1 = -51 - 1\cos 26^\circ - 109\cos 52^\circ + 48\cos 78^\circ$$

$$= -51 - 1 - 67 + 10 = -109,$$

$$7A_2 = 73 + 151\cos 52^\circ + 199\cos 104^\circ + 242\cos 156^\circ$$

$$= 73 + 93 - 48 - 221 = -103,$$

$$7B_1 = 45\sin 26^\circ + 1\sin 52^\circ + 74\sin 78^\circ$$

$$= 20 + 1 + 72 = 93,$$

$$7B_2 = -71\sin 52^\circ - 55\sin 104^\circ - 16\sin 156^\circ$$

$$= -59 - 53 - 6 = -118.$$

Using a slide rule, we find

$$\phi_1 = 139^\circ, \quad R_1 = 144 \times \tfrac{2}{14} = 21;$$

$$\phi_2 = 229^\circ, \quad R_2 = 156 \times \tfrac{2}{14} = 22.$$

The accuracy with which these two harmonics represent the observations is readily tested. The mean of the original observations is $A_0 = 47\cdot5$. Assuming 48 as the mean and correcting afterwards for the difference, we find the (standard deviation)^2 is 714. The square of the standard deviation, after correcting for the first two harmonics, is

$$714 - \tfrac{1}{2}(21^2 + 22^2) = 714 - 462 = 252.$$

Thus these two harmonics represent about two-thirds of the variance of the observations. It is thus possible that a higher harmonic may be as great as the first or second, and the representation of the observations by the two harmonics only is not completely satisfactory.

Example 3. Find a trigonometric series which shall represent accurately the following series of determinations of the monthly mean temperature on Ben Nevis during 1902:

Jan.	23°·8 F.		July	37°·6 F.
Feb.	22 ·2		Aug.	38 ·0
March	25 ·8		Sept.	38 ·4
April	27 ·1		Oct.	32 ·4
May	27 ·6		Nov.	30 ·0
June	39 ·5		Dec.	25 ·9

We have $q = 30°$, $\cos 30° = {\cdot}866$, $\cos 60° = {\cdot}5$, $\cos 90° = 0$,

$$12A_0 = 183{\cdot}2 + 185{\cdot}1 = 368{\cdot}3.$$

r	$\cos\theta$	$a' \cos\theta$	r	$\cos\theta$	$a \cos\theta$
0	1	$-13{\cdot}8$	0	1	$61{\cdot}4$
1	$\pm{\cdot}866$	$\mp 25{\cdot}46$	1	$\pm 0{\cdot}5$	$\pm 62{\cdot}8$
2	$0{\cdot}5$	$-5{\cdot}1$	2	$-0{\cdot}5$	$-60{\cdot}9$
			3	∓ 1	$\mp 59{\cdot}5$

(Upper sign) $6A_1 = -44{\cdot}36$.   (Upper sign) $6A_2 = 3{\cdot}8$.

(Lower sign) $5A_5 = 6{\cdot}56$.   (Lower sign) $6A_4 = -2{\cdot}8$.

$6A_3 = a_0' - a_2' = -3{\cdot}6$. $12A_6 = a_0 - a_1 + a_2 - a_3 = 183{\cdot}2 - 185{\cdot}1 = -1{\cdot}9$.

r	l		a	b	α	α'	β	β'
0	$23{\cdot}8$		$23{\cdot}8$		$61{\cdot}4$	$-13{\cdot}8$		
1	$22{\cdot}2$	$25{\cdot}9$	$48{\cdot}1$	$-3{\cdot}7$	$125{\cdot}6$	$-29{\cdot}4$	$-2{\cdot}2$	$-5{\cdot}2$
2	$25{\cdot}8$	$30{\cdot}0$	$55{\cdot}8$	$-4{\cdot}2$	$121{\cdot}8$	$-10{\cdot}2$	$-15{\cdot}0$	$6{\cdot}6$
3	$27{\cdot}1$	$32{\cdot}4$	$59{\cdot}5$	$-5{\cdot}3$	$59{\cdot}5$		$-5{\cdot}3$	
4	$27{\cdot}6$	$38{\cdot}4$	$66{\cdot}0$	$-10{\cdot}8$	$a_0 + a_2$ $= 183{\cdot}2$	$a_0' - a_2'$ $= -3{\cdot}6$	$\beta_1 - \beta_3$ $= 3{\cdot}1$	
5	$39{\cdot}5$	$38{\cdot}0$	$77{\cdot}5$	$1{\cdot}5$				
6	$37{\cdot}6$		$37{\cdot}6$		$a_1 + a_3$ $= 185{\cdot}1$			

For the B's,

r	$\sin\theta$	$\beta \sin\theta$	r	$\sin\theta$	$\beta' \sin\theta$
1	$0{\cdot}5$	$-1{\cdot}1$	1	$0{\cdot}866$	$-4{\cdot}50$
2	$\pm 0{\cdot}866$	$\mp 12{\cdot}99$	2	$\pm 0{\cdot}866$	$\pm 5{\cdot}71$
3	1	$-5{\cdot}3$			

(Upper sign) $6B_1 = -19{\cdot}39$.   (Upper sign) $6B_2 = 1{\cdot}21$.

(Lower sign) $6B_5 = 6{\cdot}59$.   (Lower sign) $6B_4 = -10{\cdot}21$.

$$6B_3 = \beta_1 - \beta_3 = 3{\cdot}1.$$

The required trigonometrical expression is therefore

Temperature $= 30°{\cdot}69 - 7°{\cdot}39 \cos\theta + 0°{\cdot}63 \cos 2\theta - 0°{\cdot}60 \cos 3\theta$

$$- 0°{\cdot}47 \cos 4\theta + 1°{\cdot}09 \cos 5\theta - 0°{\cdot}16 \cos 6\theta$$

$$- 3°{\cdot}23 \sin\theta + 0°{\cdot}20 \sin 2\theta + 0°{\cdot}52 \sin 3\theta - 1°{\cdot}70 \sin 4\theta$$

$$+ 1°{\cdot}10 \sin 5\theta.$$

A simple check on a portion of the work is obtained by putting $\theta = 0$ in this expression. Its value is then

$$30\cdot69 - 7\cdot39 + 0\cdot63 - 0\cdot60 - 0\cdot47 + 1\cdot09 - 0\cdot16 = 23\cdot79,$$

which compares favourably with the measured temperature for January, $23°\cdot8$.

Again, since the formula should give an exact representation of the 12 temperatures, $[vv]$ as derived from equation (11) should be zero.

$$[vv] = 11734\cdot2 - 12 \times (30\cdot69)^2 - 12 \times (0\cdot16)^2$$
$$- 6\{(7\cdot39)^2 + (0\cdot63)^2 + (0\cdot60)^2 + (0\cdot47)^2 + (1\cdot09)^2$$
$$+ (3\cdot23)^2 + (0\cdot20)^2 + (0\cdot52)^2 + (1\cdot70)^2 + (1\cdot10)^2\} = 5\cdot34.$$

This is sufficiently near to zero in view of the order of approximation to which the work has been carried out.

Example 4. The relative position of two stars is measured as follows:

Two parallel wires are made to pass each through one star, and the distances $z_0, z_1, z_2, \dots, z_{11}$ between the two wires are measured when the wires make angles of 0, 30°, 60°, 90°, ..., 330° with a fixed axis in their plane. The following 12 measures give the distances z_0, z_1, etc. in terms of one division of the micrometer head; and one division is equal to $0''\cdot4208$. Find the angular distance of the two stars.

$z_0 = 50\cdot4$,	$z_4 = 64\cdot0$,	$z_8 = 49\cdot6$,
$z_1 = 54\cdot3$,	$z_5 = 61\cdot8$,	$z_9 = 46\cdot2$,
$z_2 = 59\cdot9$,	$z_6 = 58\cdot9$,	$z_{10} = 46\cdot0$,
$z_3 = 62\cdot7$,	$z_7 = 53\cdot8$,	$z_{11} = 47\cdot2$.

It can easily be shown that z is given by

$$z_r = A_0 + A_1 \cos\theta_r + B_1 \sin\theta_r, \qquad \text{where } \theta_r = r \cdot 30°.$$

Then ρ, the angular distance between the two stars, is given by

$$\rho^2 = A_1{}^2 + B_1{}^2,$$

and we need only evaluate A_1 and B_1, following the scheme of the preceding example.

$$Ans. \quad \rho = 9\cdot174 = 3''\cdot86.$$

Example 5. The following table gives the monthly mean temperatures recorded at Greenwich from 1841 to 1890. Obtain a complete Fourier series to represent these values.

Jan.	$38°\cdot5$ F.	May	$53°\cdot1$ F.	Sept.	$57°\cdot2$ F.
Feb.	39 ·5	June	59 ·4	Oct.	50 ·0
March	41 ·7	July	62 ·5	Nov.	43 ·2
April	47 ·2	Aug.	61 ·6	Dec.	39 ·7

Example 6. Find the first four harmonics in the Fourier series for the following 24 numbers:

97 94 100 110 103 101 85 101 115 91 106 106

102 113 102 105 99 112 81 100 106 94 83 94

In order to decrease the numbers with which we have to deal, it is advisable to subtract 80 from the numbers given. This only affects the constant term, and can be allowed for later.

The numbers are written in the order suggested in §§ 85, 86.

	0	1	2	3	4	5	6	7	8	9	10	11	12
	17	14	20	30	23	21	5	21	35	11	26	26	22
		14	3	14	26	20	1	32	19	25	22	33	
a	17	28	23	44	49	41	6	53	54	36	48	59	22
b		0	17	16	−3	1	4	−11	16	−14	4	−7	

a	{17	28	23	44	49	41	6
	{22	59	48	36	54	53	

a	39	87	71	80	103	94	6
a'	−5	−31	−25	8	−5	−12	

b	{	0	17	16	−3	1	4
	{	−7	4	−14	16	−11	

β		−7	21	2	13	−10	4
β'		7	13	30	−19	12	

							1	2	3

a	{39	87	71	80
	{6	94	103	

β'	{7	13	30
	{12	−19	

γ	45	181	174	80
γ'	33	−7	−32	

ϵ	19	−6	30
ϵ'	−5	32	

$$24 A_0 = 45 + 181 + 174 + 80 = 480,$$

$$12 A_1 = -5 - 31 \cos 15° - 25 \cos 30° + 8 \cos 45°$$
$$- 5 \cos 60° - 12 \cos 75° = -56·6,$$

$$12 A_2 = 33 - 7 \cos 30° - 32 \cos 60° = 11,$$

$$12 A_3 = -5 - 31 \cos 45° - 25 \cos 90° + 8 \cos 135°$$
$$- 5 \cos 180° - 12 \cos 225° = -19·1,$$

$$12 A_4 = 45 + 181 \cos 60° + 174 \cos 120° + 80 \cos 180° = -31·5.$$

$$12 B_1 = -7 \sin 15° + 21 \sin 30° + 2 \sin 45° + 13 \sin 60°$$
$$- 10 \sin 75° + 4 \sin 90° = 15·7,$$

$$12 B_2 = 19 \sin 30° - 6 \sin 60° + 30 \sin 90° = 34·3,$$

$$12 B_3 = -7 \sin 45° + 21 \sin 90° + 2 \sin 135° + 13 \sin 180°$$
$$- 10 \sin 225° + 4 \sin 270° = 10·5,$$

$$12 B_4 = -5 \sin 60° + 32 \sin 120° = 23·4.$$

The above figures yield $A_0 = 20$. In order to correct for the original subtraction of 80 from all the figures analysed, we now add 80 to this value of A_0. The Fourier series then reads

$$100 - 4\cdot7 \cos\theta + 0\cdot9 \cos 2\theta - 1\cdot6 \cos 3\theta - 2\cdot6 \cos 4\theta$$
$$+ 1\cdot3 \sin\theta + 2\cdot8 \sin 2\theta - 0\cdot9 \sin 3\theta + 2\cdot0 \sin 4\theta,$$

where $\theta = 0$, 15°, 30°, etc. for the first, second, and third figures respectively.

Example 7. Find the first harmonic term in the Fourier series representing the following 20 quantities:

88	124	103	90	49	168	99	62	104	97
51	−28	−1	42	55	59	99	38	122	37

Since only the coefficients A_1 and B_1 are required, the work may be considerably shortened.

	0	1	2	3	4	5	6	7	8	9	10
	88	124	103	90	49	168	99	62	104	97	51
		37	122	38	99	59	55	42	−1	−28	
a	{ 88	161	225	128	148	227	154	104	103	69	51
	{ 51	69	103	104	154						
a' (Subtract)	37	92	122	24	−6						
b		{ 87	−19	52	−50	109	[44	20	105	125]	
		{ 125	105	20	44						
β (Add)		212	86	72	−6	109					

$10 A_1 = 37 + 92 \cos 18° + 122 \cos 36° + 24 \cos 54° - 6 \cos 72° = 235\cdot3,$

$10 B_1 = 212 \sin 18° + 86 \sin 36° + 72 \sin 54° - 6 \sin 72° + 109 \sin 90° = 277\cdot5.$

Hence the required terms are

$$23\cdot5 \cos\theta + 27\cdot7 \sin\theta.$$

90. Practical Method of investigating Periodicities.

In practice, we should not endeavour to find the amplitude of a periodic term from observations extending over one period only. Thus a periodic term of one year would only be investigated by means of observations extending over a large number of years. The work is then arranged as follows.

Let u_0, u_1, u_2, etc. be a series of observations taken at equal intervals of time a; and suppose it is required to investigate

XI]   HARMONIC ANALYSIS   199

a possible period $p\alpha$. The u's are written in rows of p each, as follows:

$$
\begin{array}{llll}
u_0 & u_1 & u_2 & \ldots\; u_{p-1} \\
u_p & u_{p+1} & u_{p+2} & \ldots\; u_{2p-1} \\
u_{2p} & u_{2p+1} & u_{2p+2} & \ldots\; u_{3p-1} \\
\multicolumn{4}{c}{\dotfill} \\
\multicolumn{4}{c}{\dotfill} \\
\multicolumn{4}{c}{\dotfill} \\
u_{(n-1)p} & \multicolumn{3}{l}{\dotfill\; u_{np-1}}
\end{array}
$$

$$
\overline{\quad V_0 \quad\quad V_1 \quad\quad V_2 \quad\quad\quad V_{p-1} \quad}
$$

We shall suppose that there are sufficient observations to form n rows of p each. The columns are added, yielding the sums $V_0, V_1, \ldots, V_{p-1}$. The sums V_0, V_1, etc. are then analysed harmonically by the methods outlined above, yielding the expression

$$V = A_0 + A_1 \cos\theta + B_1 \sin\theta + \ldots,$$

where V takes the value V_r when $\theta = \dfrac{2\pi r}{p}$,

$$A_1 = \frac{2}{p}\sum_{r=0}^{p-1} V_r \cos\frac{2\pi r}{p} = \frac{2}{p}\sum_{r=0}^{pn-1} u_r \cos\frac{2\pi r}{p},$$

$$B_1 = \frac{2}{p}\sum_{r=0}^{p-1} V_r \sin\frac{2\pi r}{p} = \frac{2}{p}\sum_{r=0}^{pn-1} u_r \cos\frac{2\pi r}{p},$$

$$A_0 = \frac{1}{p}\sum_{r=0}^{p-1} V_r.$$

When it is only required to investigate a simple periodicity of period p intervals, it is not necessary to carry the Fourier expansion beyond the terms in $\cos\theta$ and $\sin\theta$.

The amplitude in the V's of the period $p\alpha$ is

$$(A_1{}^2 + B_1{}^2)^{\frac{1}{2}}.$$

The amplitude of the period $p\alpha$ in the original observations u_0, u_1, etc. will be

$$\frac{1}{n}(A_1{}^2 + B_1{}^2)^{\frac{1}{2}}.$$

For if $t = a_0 + a_1 \cos\theta + b_1 \sin\theta + \ldots,$

where t takes the value u_r for $\theta = \dfrac{2\pi r}{p}$, then each of the periodic

terms is repeated exactly down the columns of the table of
u's. The amplitude of the variation in the V's is thus n times
the amplitude of the same period in the u's. Thus we may
write

$$(a_1{}^2 + b_1{}^2)^{\frac{1}{2}} = \frac{1}{n}(A_1{}^2 + B_1{}^2)^{\frac{1}{2}}.$$

The right-hand side of this equation gives the required
amplitude.

The method has the advantage of destroying to a certain
extent accidental errors in individual u's. Periods of p intervals
or of any exact sub-multiple of this period will be intensified n-fold
in the sequence V_0, V_1, V_2, etc.; while other periodicities will
tend to be destroyed by the process of addition. For the latter
periodicities will occur with different phases in the horizontal
rows of the table of u's, and the terms so produced will tend to
annul one another on addition of the columns, provided n is large,
and the periods concerned be not too nearly equal to the sub-
periods of p intervals.

The figures given in Ex. 5, p. 196, afford an example of this
method. The temperature of $38°\cdot5$ given for January is the mean
of the January temperature for 50 years. This corresponds to the
method shown in the table on page 199, except that the V's are
all divided by n, the result being treated as the mean annual
variation. In practice it will sometimes be found convenient to
analyse the quantities V_0, V_1, etc. as they stand, while in other
cases it will be found preferable to divide the V's by n, treating the
result as the mean variation of period pa.

Further, it should be noted that if the same quantity be added
to, or subtracted from, all the terms which we desire to analyse
harmonically, only the constant A_0 is affected, the periodic terms
being unaffected. It is sometimes useful to remember this property, as the numbers with which we have to deal can often be
reduced in magnitude by subtracting the same amount from all
the terms. Thus in Ex. 5, p. 196, we should obtain the same
periodic terms if we subtracted $38°\cdot5$ from all the terms before
proceeding further.

Tables for facilitating the evaluation of $\Sigma\, V_r \cos r\theta$, $\Sigma\, V_r \sin r\theta$ have been
drawn up by H. H. Turner, and published under the title *Tables for
Facilitating the Use of Harmonic Analysis* (Oxford Univ. Press).

A large number of instruments and graphical methods have been devised for performing the evaluation of the coefficients of the terms of the Fourier series for any given set of observations. For a description of these instruments and methods, as well as of other instruments, tables, etc., the reader is referred to the Handbook of the Napier Tercentenary Celebration of 1914, published by Messrs G. Bell and Sons, under the title of *Modern Instruments of Calculation.* This work is a mine of information concerning calculating instruments of all kinds.

91. Correction of the Amplitudes for the effect of using mean values for the computation.

If the original data l_0, l_1, etc. are mean values of some variable quantity, the methods described above will yield the best representation of the mean values, and the computed amplitudes will refer to variations in these means. Suppose for example the original data are monthly mean temperatures, it is obvious that the process of evaluating monthly means will partially smooth out the variations in temperature, and we require to obtain an estimate of the extent of this smoothing.

Let the variations of temperature be represented by the series

$$L = A_0' + A_1' \cos t + A_2' \cos 2t + \dots$$
$$+ B_1' \sin t + B_2' \sin 2t + \dots.$$

We can now use this relation to form the mean values for any desired interval. Let there be n such intervals between $t = 0$ and $t = 2\pi$, so that we shall form n mean values $l_0, l_1, \dots l_{n-1}$, and l_s will be the mean value of L through the interval $(s - \frac{1}{2})\dfrac{2\pi}{n}$ and $(s + \frac{1}{2})\dfrac{2\pi}{n}$.

Then

$$l_s = \frac{n}{2\pi} \int_{(s-\frac{1}{2})\frac{2\pi}{n}}^{(s+\frac{1}{2})\frac{2\pi}{n}} L \, dt$$

$$= A_0' + \frac{n}{2\pi} \Sigma A_r' \frac{1}{r} \left\{ \sin r \left(s + \frac{1}{2}\right) \frac{2\pi}{n} - \sin r \left(s - \frac{1}{2}\right) \frac{2\pi}{n} \right\}$$

$$+ \frac{n}{2\pi} \Sigma B_r' \frac{1}{r} \left\{ \cos r \left(s - \frac{1}{2}\right) \frac{2\pi}{n} - \cos r \left(s + \frac{1}{2}\right) \frac{2\pi}{n} \right\}$$

$$= A_0' + \Sigma A_r' \frac{\sin \dfrac{r\pi}{n}}{\dfrac{r\pi}{n}} \cos \frac{rs \, 2\pi}{n} + \Sigma B_r' \frac{\sin \dfrac{r\pi}{n}}{\dfrac{r\pi}{n}} \sin \frac{rs \, 2\pi}{n}.$$

But this is equivalent to the series

$$l_s = A_0 + \sum_{r=1} A_r \cos rt_s + \sum_{r=1} B_r \sin rt_s,$$

where

$$t_s = \frac{s\,2\pi}{n}.$$

Hence the amplitude of the true variation in the variable L is derived from the amplitude computed from the mean values by multiplying the latter by

$$\frac{r\pi/n}{\sin\,r\pi/n}.$$

92. Sine and Cosine Series.

Fourier's Theorem as stated in § 81 above leads to the representation of a function $f(\theta)$ by a series of sines and cosines within the range $0 \leqslant \theta \leqslant 2\pi$. It is sometimes required, however, to represent a function $f(\theta)$ by a series of sine terms only, or of cosine terms only, within the range $0 \leqslant \theta \leqslant \pi$.

Cosine Series. If we define a function $F(\theta)$ so that

$$F(\theta) = f(\theta) \qquad 0 \leqslant \theta \leqslant \pi$$
$$F(\theta) = f(2\pi - \theta) \qquad \pi \leqslant \theta \leqslant 2\pi,$$

the function $F(\theta)$ is everywhere continuous, provided $f(\theta)$ is continuous from 0 to π. It is therefore possible to represent $F(\theta)$ by a Fourier series, which will therefore represent $f(\theta)$ within the range 0 to π.

$$F(\theta) = A_0 + A_1 \cos \theta + A_2 \cos 2\theta + \dots$$
$$+ B_1 \sin \theta + B_2 \sin 2\theta + \dots.$$

The coefficient A_s is given by equation (3), page 179.

$$\pi A_s = \int_0^{2\pi} F(\theta) \cos s\theta\, d\theta$$

$$= \int_0^{\pi} f(\theta) \cos s\theta\, d\theta + \int_\pi^{2\pi} f(2\pi - \theta) \cos s\theta\, d\theta.$$

In the second integral change the variable to θ' where $\theta' = 2\pi - \theta$.

$$\pi A_s = \int_0^{\pi} f(\theta) \cos s\theta\, d\theta + \int_0^{\pi} f(\theta') \cos s\theta'\, d\theta'$$

$$= 2\int_0^{\pi} f(\theta) \cos s\theta\, d\theta.$$

Similarly
$$\pi B_s = \int_0^{2\pi} F(\theta)\sin s\theta \, d\theta$$
$$= \int_0^{\pi} f(\theta)\sin s\theta \, d\theta - \int_0^{\pi} f(\theta')\sin s\theta' \, d\theta'$$
$$= 0.$$

Hence within the range $0 \leqslant \theta \leqslant \pi$,
$$f(\theta) = A_0 + A_1 \cos\theta + A_2\cos 2\theta + \ldots$$

where
$$A_s = \frac{2}{\pi}\int_0^{\pi} f(\theta)\cos s\theta \, d\theta, \quad A_0 = \frac{1}{\pi}\int_0^{\pi} f(\theta)\, d\theta.$$

We thus obtain a series of cosine terms only, to represent the function $f(\theta)$ within the range from 0 to π.

Sine Series. If we define a function $G(\theta)$ so that
$$G(\theta) = f(\theta) \quad 0 \leqslant \theta < \pi$$
$$G(\theta) = -f(2\pi - \theta) \quad \pi < \theta \leqslant 2\pi,$$

the function $G(\theta)$ has a finite discontinuity at $\theta = \pi$. The function $G(\theta)$ may therefore be represented by a Fourier series,
$$G(\theta) = A_0 + A_1\cos\theta + A_2\cos 2\theta + \ldots$$
$$+ B_1\sin\theta + B_2\sin 2\theta + \ldots.$$

The computation of A_s and B_s is similar to that carried out above for $F(\theta)$, the sole difference being that the sign is changed before the second integral in the equations. It follows that all the A's vanish, while the B's are defined by
$$B_s = \frac{2}{\pi}\int_0^{\pi} f(\theta)\sin s\theta \, d\theta.$$

This series vanishes at $\theta = 0$ and at $\theta = \pi$, and the useful application of the sine-series is limited to cases in which it is known that $f(\theta)$ vanishes at these limits.

Modification for a finite number of observations. If we have n observations corresponding to $\theta = \dfrac{r\pi}{n}$, where
$$r = 0, 1, 2, \ldots (n-1),$$

which we shall call $l_0, l_1, l_2, \ldots l_{n-1}$, the observations may be represented by a series

$$l = A_0 + A_1 \cos \theta + A_2 \cos 2\theta + \ldots,$$

where $A_0 = \dfrac{1}{n} \Sigma\, l, \qquad A_s = \dfrac{2}{n} \Sigma\, l \cos s\theta,$

or by a series

$$l = B_1 \sin \theta + B_2 \sin 2\theta + \ldots,$$

where $B_s = \dfrac{2}{n} \Sigma\, l \sin s\theta.$

These equations for the A's or B's are readily deduced from the integral forms, as was done in § 82 above.

93. Probable Error of the Fourier Coefficients.

If each of a series of n observations is known to be subject to a known probable error ϵ, the probable error of the constant term A_0 is $\dfrac{\epsilon}{\sqrt{n}}$, since A_0 is the mean of the observations. Also since

$$A_s = \frac{2}{n} \sum_{r=0}^{r=n-1} l_r \cos r\theta_s,$$

the probable error of $A_s = \dfrac{2}{n}\, \epsilon (\Sigma \cos^2 r\theta_s)^{\frac{1}{2}} = \dfrac{2}{n}\, \epsilon \sqrt{\dfrac{n}{2}} = \epsilon \sqrt{\dfrac{2}{n}}.$

Again, since

$$B_s = \frac{2}{n} \sum_{r=1}^{r=n-1} l_r \sin r\theta_s,$$

the probable error of

$$B_s = \frac{2}{n}\, \epsilon\, (\Sigma \sin^2 r\theta_s)^{\frac{1}{2}} = \epsilon \sqrt{\frac{2}{n}}.$$

If $R_s = \sqrt{A_s^2 + B_s^2},$

$$dR_s = \frac{A_s\, dA_s + B_s\, dB_s}{\sqrt{A_s^2 + B_s^2}}.$$

Hence, probable error of $R_s = \epsilon \sqrt{\dfrac{2}{n}}.$

If the standard deviation of the original observations is σ, the standard deviation of A_0, the mean, is $\dfrac{\sigma}{\sqrt{n}}$ and the standard deviation of any other coefficient A or B and of the amplitude R is $\dfrac{\sigma}{\sqrt{\frac{1}{2}n}}$.

Also since
$$\phi = \tan^{-1}\frac{B}{A},$$

$$d\phi = \frac{A\,dB - B\,dA}{A^2 + B^2},$$

and the standard deviation of $\phi = \dfrac{\sigma}{\sqrt{\frac{1}{2}n}} \times \dfrac{1}{\sqrt{a^2 + b^2}}.$

CHAPTER XII

94. Hidden Periodicities.

In certain classes of observations, the length of the main period becomes obvious as soon as the observations are examined; e.g. the period of the semi-diurnal tide can be deduced from a comparatively small number of observations. Once the length of the period is known, the methods of the preceding chapter can be immediately applied to deduce the amplitude and phase. But when the length of the period is unknown, and cannot be deduced in a simple manner, the difficulties of the investigation are enormously increased. Thus, if we consider records of temperature extending over a large number of years, we shall find that, once the effect of the annual period and its harmonics has been removed, the resulting records show no obvious periodicity, though they are probably due to a number of superposed periodicities, with certain accidental variations added on. In meteorological phenomena generally, the changes from day to day appear to be so arbitrary, that one is forced to the conclusion that, whatever periodicities may underlie the phenomena, they will be very largely masked by apparently accidental variations. The methods considered in the present chapter aim at unmasking such underlying periodicities, and determining their amplitudes and phases.

It has already been shown in Chapter XI that it is always possible to find a Fourier series which shall represent with any required degree of accuracy any given set of numbers; but it must not be assumed that the Fourier series is always an accurate representation of the physical laws underlying the phenomena. For it is obvious that even a number of quantities distributed at random can be represented by a Fourier series; though in such a case the use of the series might lead to the false impression that the phenomena under consideration were due to the combined action of a number of purely periodic physical causes. The real difficulty lies, not so much in finding a Fourier series to fit the observations, as in determining which of a large number of possible periods have sufficiently great amplitudes to be regarded as having real physical significance.

95. The Periodogram.

The Periodogram method of searching for periodicities consists essentially in finding (by the method of § 90, or otherwise) the amplitudes of a large number of trial periods. The trial periods which yield the greatest amplitudes will yield approximations to the most probable periodicities. It will be proved later that if a trial period T should fall near one of the actual periods of the observations, the resulting amplitude R yielded by the method of § 90 will be considerably greater than if the trial period T were considerably removed from the actual periods. Thus if R^2 be plotted for different values of T, the points of maxima of the curve will yield the most likely periods. If N observations be used in deriving R for the period T, it will be found in some ways preferable to plot $R^2 N$ for different values of T; but in practice if we had say 600 readings to hand for a periodogram investigation we should use all or practically all of the readings for investigating all trial periods. Thus N will never in practice vary considerably, and it is sufficient to plot R^2 for different values of T. The rather ill-defined maxima of this curve will yield approximations to the most likely periods. More accurate values of these periods will afterwards be derived by what is called the Secondary Analysis (§ 97). Finally it is necessary to consider how great the amplitude of any of these periods must be, in order that we may be certain that it did not arise from a purely chance distribution of the quantities considered.

It is of course impracticable to find R^2 for all possible values of T, and Schuster has shown that, if a period T has been investigated, the nearest period to this which need be investigated is T', given by the equation

$$n(T - T') = \pm \tfrac{1}{4} T.$$

This limit is set by the fact that any two periods which may be investigated should be independent, and two near periods will begin to be independent when there is a final difference of phase of a quarter-period. This condition leads to the above equation, where n is the number of periods T used in the investigation.

96. Form of the Periodogram for one simple period.

Suppose the observations to be analysed could be represented by a single periodic function $R' \cos(\kappa t - \epsilon)$, where $\kappa = \dfrac{2\pi}{T'}$, T' being the true period of the series. Let T be one of the primary periods for which the observations have been analysed, and let $T = \dfrac{2\pi}{g}$. Then if the number of observations be large, the summations used in the usual harmonic analysis can be replaced by integrations. The coefficients of the first two terms in the Fourier sequence then become [*]

$$\tfrac{1}{2}nTA_1 = R' \int_0^{nT} \cos(\kappa t - \epsilon) \cos gt\, dt,$$

$$\tfrac{1}{2}nTB_1 = R' \int_0^{nT} \cos(\kappa t - \epsilon) \sin gt\, dt,$$

where n is the total number of complete periods T used in the analysis. Integrating these expressions, and remembering that

$$gnT = 2\pi n,$$

we obtain

$$\tfrac{1}{2}nTA_1 = \tfrac{1}{2}R' \int_0^{nT} \{\cos(\overline{\kappa + g}\,t - \epsilon) + \cos(\overline{\kappa - g}\,t - \epsilon)\}\, dt$$

$$= R' \frac{2\kappa}{\kappa^2 - g^2} \sin \tfrac{1}{2}\kappa nT \cos(\tfrac{1}{2}\kappa nT - \epsilon),$$

$$\tfrac{1}{2}nTB_1 = - R' \frac{2g}{\kappa^2 - g^2} \sin \tfrac{1}{2}\kappa nT \sin(\tfrac{1}{2}\kappa nT - \epsilon).$$

Or, if R be the amplitude of the period T yielded by the primary analysis,

$$\frac{R}{R'} = \frac{4}{(\kappa^2 - g^2)\, nT} \sin \tfrac{1}{2}\kappa nT \{\kappa^2 \cos^2(\tfrac{1}{2}\kappa nT - \epsilon) + g^2 \sin^2(\tfrac{1}{2}\kappa nT - \epsilon)\}^{\frac{1}{2}}.$$

Putting $\gamma = \tfrac{1}{2}(\kappa - g)\, nT = \pi n\left(\dfrac{\kappa}{g} - 1\right),$

[*] Cf. equations (12), page 182.

and

$$\sin \tfrac{1}{2}\kappa nT = \sin \{\tfrac{1}{2}(\kappa - g)\, nT + \tfrac{1}{2}gnT\} = \sin (\gamma + n\pi) = (-1)^n \sin \gamma,$$

we obtain the relation

$$\frac{R}{R'} = \frac{\sin \gamma}{\gamma} \times \frac{2\{\kappa^2 \cos^2 (2\gamma - \epsilon) + g^2 \sin^2 (2\gamma - \epsilon)\}^{\frac{1}{2}}}{\kappa + g}.$$

When κ and g are equal or very nearly equal, this reduces to

$$\frac{R}{R'} = \frac{\sin \gamma}{\gamma}.$$

Since

$$\gamma = \tfrac{1}{2}(\kappa - g)\, nT = \pi n \frac{\kappa - g}{\kappa} = \tfrac{1}{2}\kappa nT - \pi n = \pi n \left(\frac{T}{T'} - 1\right),$$

it follows that $\dfrac{\sin \gamma}{\gamma}$ only has appreciable values when $\kappa - g$ is small. For the function $\dfrac{\sin \gamma}{\gamma}$ has its maximum at $\gamma = 0$, and decreases to zero at $\gamma = \pi$, after which it has a succession of maxima and minima whose amplitudes are small.

Considering the equation

$$\frac{R^2}{R'^2} = \frac{\sin^2 \gamma}{\gamma^2} \times \frac{4\{\kappa^2 \cos^2 (2\gamma - \epsilon) + g^2 \sin^2 (2\gamma - \epsilon)\}}{(\kappa + g)^2},$$

we see that the second factor only varies slowly, so that the curve representing the values of $\dfrac{R^2}{R'^2}$ for different values of γ is not appreciably altered, so far as its general form is concerned, when this factor is neglected. The general distribution of $\dfrac{R^2}{R'^2}$ is approximately of the form of $\dfrac{R^2}{R'^2} = \dfrac{\sin^2 \gamma}{\gamma^2}$ at points near $\gamma = 0$. The curve will show a broad band at $\gamma = 0$, and on each side of this band a succession of other bands of rapidly decreasing intensity. The result is analogous with the formation of diffraction bands (fig. 12). The greater the total number n of periods used, the nearer will the diffraction bands crowd together on each side of the principal band, and the narrower will the principal band be.

It is thus seen that a curve showing R^2 for different values of γ, which is of precisely the same form as the curve representing R^2 for different values of T $\left[\text{since } \gamma = \pi n \left(\dfrac{T}{T'} - 1\right)\right]$, shows a broad band at $\gamma = 0$ or $T = T'$, and a succession of subsidiary bands of small maximum ordinate on each side of this band. Since it was assumed in the first place that the period T' was the only true period, the curve represents the effect of the true period T' upon

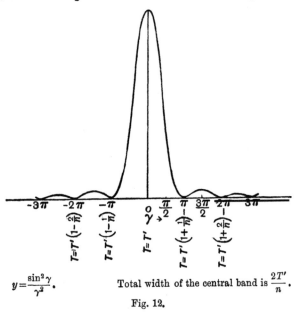

$$y = \frac{\sin^2 \gamma}{\gamma^2}.$$

Total width of the central band is $\dfrac{2T'}{n}$.

Fig. 12.

the square of the amplitude of other trial periods analysed by the periodogram; and it has been shown that this effect is negligible except for such trial periods T as will yield small values of γ. The central band does not spread beyond

$$\gamma = \pm \pi, \text{ or } n\left(\frac{T}{T'} - 1\right) = \pm 1.$$

The limits of the central band are therefore

$$T = T'\left(1 \pm \frac{1}{n}\right),$$

where n is the total number of complete periods considered in the investigation.

Thus we were justified in the assumption made in § 89 that the periodogram spreads a periodicity into a band. It has been shown also that the width of the band can be decreased by increasing the number of complete periods considered in the analysis. But since $\dfrac{\sin^2 \gamma}{\gamma^2}$ only changes slowly in the neighbourhood of $\gamma = 0$, the maximum shown in the periodogram will be flat, and the actual point of maximum will therefore be ill-defined. It must therefore not be expected that the primary analysis of the periodogram should yield the exact value of the period.

97. Secondary Analysis.

When the periodogram has shown the presence of a true periodicity in the neighbourhood of a primary period T, the true period T' can be obtained by analysing harmonically each period T, or the sums of groups of m periods. It will be supposed that the number of observations is sufficiently large to permit of our replacing summations by integrals.

In the first place suppose the observations u_0, u_1, etc. can be accurately represented by a simple period

$$u = C + R' \cos (\kappa t - \epsilon).$$

Let the n complete periods be divided into groups of m periods, and let the first two terms in the harmonic analysis of the sth group be

$$A_s \cos gt + B_s \sin gt, \quad \text{where} \quad g = \frac{2\pi}{T}.$$

Then

$$\tfrac{1}{2} mTA_s = \int_{(s-1)mT}^{smT} u \cos gt \, dt = R' \int_{(s-1)mT}^{smT} \cos (\kappa t - \epsilon) \cos gt \, dt$$

$$= \tfrac{1}{2} R' \int_{(s-1)mT}^{smT} \{\cos (\overline{\kappa + g}\, t - \epsilon) + \cos (\overline{\kappa - g}\, t - \epsilon)\} \, dt,$$

$$\frac{mTA_s}{R'} = \left(\frac{1}{\kappa + g} + \frac{1}{\kappa - g} \right) \sin (s\kappa mT - \epsilon)$$

$$- \left(\frac{1}{\kappa + g} + \frac{1}{\kappa - g} \right) \sin (\overline{s - 1}\, \kappa mT - \epsilon)$$

$$= \frac{4\kappa}{\kappa^2 - g^2} \sin \tfrac{1}{2} \kappa mT \cos \{\tfrac{1}{2} (2s - 1) \kappa mT - \epsilon\}.$$

14-2

Similarly

$$\frac{mTB_s}{R'} = \frac{-4g}{\kappa^2 - g^2} \sin \tfrac{1}{2}\kappa mT \sin \{\tfrac{1}{2}(2s-1)\kappa mT - \epsilon\}.$$

If ϕ_s be the phase of maximum in this group

$$\tan \phi_s = -\frac{g}{\kappa}\tan\{\tfrac{1}{2}(2s-1)\kappa mT - \epsilon\}.$$

It has already been shown that if T' be the point of maximum of the band in the periodogram, T for different points on the band will lie between $T'\left(1 \pm \dfrac{1}{n}\right)$. It has been supposed that the periodogram has shown that T' lies near T, or that T is well within the limits $T'\left(1 \pm \dfrac{1}{n}\right)$. Hence it follows that since $\dfrac{g}{\kappa} = \dfrac{T'}{T}$, this fraction will differ from unity by less than $\dfrac{1}{n}$. Since n will be fairly large, we may write unity instead of $\dfrac{g}{\kappa}$ in the above equation. Then

$$\tan \phi_s = -\tan\{\tfrac{1}{2}(2s-1)\kappa mT - \epsilon\},$$

or
$$\phi_s = 2\pi ms - \tfrac{1}{2}(2s-1)\kappa mT + \epsilon.$$

This is a linear function of s, and therefore ϕ_s should increase uniformly from group to group. If β be the increase of phase per group, then

$$\beta = 2\pi m - \kappa mT$$

$$= 2\pi m\left(1 - \frac{T}{T'}\right),$$

or
$$T' = \frac{T}{1 - \dfrac{\beta}{2\pi m}}.$$

Hence if the progression of phase per group can be determined when the observations are grouped m periods of T together, and the sum of the groups analysed, the true period can be deduced accurately.

In practice the work is conducted as follows. It is supposed that the primary analysis has shown the existence of a true periodicity near the trial period T, where T is equal to p of the intervals between successive observations. The available observations are arranged in a table containing n rows and

p columns, as in Chap. XI, § 90. The n rows are divided up into groups of m each, and the sums of the columns in each group written down. These sums are then analysed harmonically, either by the methods of § 87 or by the use of a harmonic analyser. Let the first two terms in the Fourier sequence for the sth row be

$$A_s \cos \theta + B_s \sin \theta.$$

Then the phase ϕ_s of this row is evaluated by means of the formula

$$\tan \phi_s = \frac{B_s}{A_s}.$$

The phase ϕ_s is evaluated for all the groups. The progression from group to group, β, can then be immediately evaluated. The method of procedure may perhaps be seen most easily by the consideration of a simple example.

The primary analysis having shown the existence of a period of 19 to 20 months in certain temperature records, it was required to find a more accurate estimate of the length of the period. As the records conveniently available extended over about 35 years, or 420 months, the temperatures were written down in periods of 20 months, forming a table of 21 rows. The rows were then arranged in groups of three periods and added, yielding seven groups of three periods each. The sums for each group were then analysed as far as the first harmonic term only. The results are given in the following table:

Group	A_s	B_s	$\dfrac{B_s}{A_s}$	ϕ_s
1	235·3	277·5	1·18	50°
2	367·8	− 10·0	− ·027	−1·5°
3	371·4	53·4	·144	8°
4	− 44·6	−124·8	2·79	250°
5	−206·1	−116·7	·57	209°
6	140·8	−177·8	−1·27	−52°
7	−327·3	87·1	− ·27	165°

The progression of phase from group to group is not obvious from this table, but may be evaluated by the following graphical method (see fig. 13). For each group plot the phase of maximum,

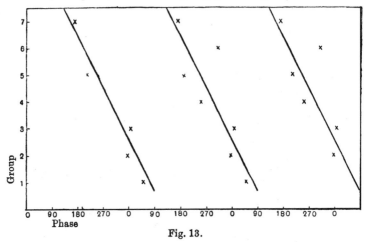

Fig. 13.

remembering that the addition of any multiple of $\pm 360°$ to the phase makes no difference to the trigonometrical expression. The points in the diagram group themselves about a number of parallel straight lines, whose slope corresponds to a decrease of phase of $36°$ from one group to the next. The corrected period is therefore

$$\frac{20}{1 + \dfrac{36}{3 \times 360}} \text{ months} = \frac{600}{31} \text{ months} = 19\cdot35 \text{ months.}$$

98. Fourier Series for a Series of Observations following the Normal Law.

The observations $l_0, l_1, \dots l_{n-1}$ will be regarded as a sample of n observations from a very large number N of quantities following the Normal Error Law, with a parameter h. The n observations are analysed by the methods shown above, giving for the sth harmonic

$$A_s \cos s\theta + B_s \sin s\theta \quad \text{or} \quad R_s \cos (s\theta - \phi_s).$$

The problem which we have to solve is, What will be the distribution of the computed R's, and what is the probability that for any s taken at random R_s shall exceed certain specified limits?

Since the A's and B's are linear functions of the l's, they will also follow the normal error law, in accordance with § 15 above.

$$A_s = \frac{2}{n} \sum_{r=0}^{r=n-1} l_r \cos s\theta_r,$$

$$B_s = \frac{2}{n} \sum_{r=0}^{r=n-1} l_r \sin s\theta_r,$$

where $\theta_r = 2\pi r/n$.

The probability that A_s lies between A and $A + dA$ is

$$\frac{H}{\sqrt{\pi}} e^{-H^2 A^2} dA,$$

where

$$\frac{1}{H^2} = \frac{4}{n^2} \sum_{r=0}^{r=n-1} \frac{1}{h^2} \cos^2 2\pi rs/n = \frac{2}{nh^2}.$$

The probability that A_s lies between A and $A + dA$ is therefore

$$\sqrt{\frac{n}{2\pi}} h e^{-\frac{1}{2}nh^2 A^2} dA.$$

The probability that B_s lies between B and $B + dB$ is similarly

$$\sqrt{\frac{n}{2\pi}} h e^{-\frac{1}{2}nh^2 B^2} dB.$$

The probability that A_s shall lie between A and $A + dA$ while B_s lies between B and $B + dB$ is

$$\frac{n}{2\pi} h^2 e^{-\frac{1}{2}nh^2 (A^2 + B^2)} dA\, dB.$$

But $A^2 + B^2 = R^2$, and $dA\, dB$ represents a small rectangular element of area in the AB plane. In order to determine the probability that R_s shall lie between R and $R + dR$ we integrate the last expression over the area between circles centred at the origin, whose radii are R and $R + dR$. The resulting probability is

$$\frac{n}{2\pi} h^2 e^{-\frac{1}{2}nh^2 R^2} 2\pi R\, dR = \frac{1}{2}nh^2 e^{-\frac{1}{2}nh^2 R^2} dR^2.$$

From this we find by integration that the probability that R_s^2 shall exceed R^2 is

$$\int_R^\infty \frac{1}{2}nh^2 e^{-\frac{1}{2}nh^2 R^2} dR^2 = e^{-\frac{1}{2}nh^2 R^2}.$$

The mean value of R_s^2 is

$$\int_0^\infty \frac{1}{2}nh^2 e^{-\frac{1}{2}nh^2 R^2} R^2\, dR^2 = \frac{2}{nh^2} \int_0^\infty z e^{-z} dz = \frac{2}{nh^2}.$$

It is convenient to call this mean value $\overline{R^2}$. Then the probability that $R_s{}^2$ shall exceed $\kappa \overline{R^2}$ is seen to be $e^{-\kappa}$.

This is the criterion originally derived by Schuster, and it has been used by all subsequent writers as the basis of discussion of the results of harmonic analysis. The interpretation of Schuster's criterion requires careful definition. In the derivation of the formulae above, we considered the distribution of the magnitude of the amplitude R_s, where the order of the harmonic, s, was chosen at random. It is not justifiable to select a harmonic for which computation has yielded a large amplitude, and to apply the criterion to this amplitude alone. If for example 10 values of R^2 have been computed, the probability that among 10 one shall exceed $\kappa \overline{R^2}$ is obviously not $e^{-\kappa}$.

Attention was first directed to this point by Sir Gilbert Walker.[*] His argument proceeds as follows. There are $\dfrac{n}{2}$ independent Fourier terms[†] in the complete representation of n observations. The probability that any harmonic selected at random shall yield a value of $\overline{R^2}$ in excess of κR^2 is $e^{-\kappa}$. The probability that R^2 shall be less than κR^2 is $1 - e^{-\kappa}$, and the probability that all $\dfrac{n}{2}$ values of R^2 shall be less than $\kappa \overline{R^2}$ is $(1 - e^{-\kappa})^{\frac{n}{2}}$. The probability that at least one R^2 shall exceed this limit is

$$1 - (1 - e^{-\kappa})^{\frac{n}{2}} = P \text{ say,}$$
$$1 - P = (1 - e^{-\kappa})^{\frac{n}{2}}.$$

If we put $P = \frac{1}{2}$, we can then derive from this equation the "probable" greatest value of R^2, which is the greatest value of R^2 which has an even chance of occurring in a series of $\dfrac{n}{2}$ terms. The "probable greatest" is therefore given by

$$1 - e^{-\kappa} = (\tfrac{1}{2})^{\frac{2}{n}},$$
$$e^{-\kappa} = 1 - (\tfrac{1}{2})^{\frac{2}{n}}.$$

The following table gives for a few values of n the corresponding values of κ.

---

* Calcutta, *Indian Met. Memoirs*, 21, 1914.

† n is large in all practical problems, and we can replace $\dfrac{n-1}{2}$ or $\dfrac{n}{2} - 1$ by $\dfrac{n}{2}$.

This table shows that if a series of 1000 terms be analysed, the "probable greatest" amplitude, assuming that the observations are distributed at random, is $\sqrt{6\cdot6}$ times the mean square amplitude, though the probability of obtaining such an amplitude for any harmonic selected at random is only ·0014, or 1 in 700.

n	κ	$e^{-\kappa}$
50	3·6	·027
100	4·3	·0138
200	5·0	·0069
500	5·9	·0028
1000	6·6	·0014
2000	7·3	·00069

It should be noted that the sum of the squares of all the amplitudes in the Fourier series is $2\sigma^2$, so that the mean R^2 computed for a given series of observations is $\dfrac{4\sigma^2}{n-1}$ if n is odd.

99. Practical Application of the Criterion.

In the practical application of the above results, the parameter h is not known *a priori*, nor is it known from what distribution the n observations form a sample. But since it is desired to compare the results of harmonic analysis of the observations with the results derivable from chance observations, the value of σ^2 is evaluated for the sample, and the value of h can be written down by the use of the equations given on page 32. The best estimate of the standard deviation of the population of which the n observations form a sample is $\sigma\sqrt{\dfrac{n}{n-1}}$. The mean value of R^2 for any harmonic is

$$\frac{\displaystyle\int_0^\infty \tfrac{1}{2}nh^2 e^{-\frac{1}{2}nh^2R^2} R^2\, dR^2}{\displaystyle\int \tfrac{1}{2}nh^2 e^{-\frac{1}{2}nh^2R^2}\, dR^2} = \frac{2}{nh^2} = \frac{4\sigma^2}{n-1},$$

since

$$h\sigma = \sqrt{\frac{n-1}{2n}}.$$

As stated this result means that the mean value of R_s^2 for a given s, for a large number of samples of n observations, will be $\dfrac{4\sigma^2}{n-1}$, if the observations are distributed at random. If, then, on analysing a series of observations we find values of R^2 not much in excess of $\dfrac{4\sigma^2}{n-1}$, we should be entitled to regard the harmonic terms as having no physical significance. If the observations were completely represented by one harmonic term, we should obtain an amplitude $\dfrac{\sigma}{\sqrt{2}}$. In practice we usually find that some amplitudes are derived which fall between the limits $\dfrac{\sigma}{\sqrt{2}}$ and $\dfrac{2\sigma}{\sqrt{n-1}}$, and the question is, Where shall the limit of physical significance be drawn? Walker's method of procedure will give a limit $\dfrac{2\sqrt{\kappa}\sigma}{\sqrt{n-1}}$, where κ is a function of n tabulated above or derived from the equation

$$e^{-\kappa} = 1 - (\tfrac{1}{2})^{\frac{2}{n}}.$$

Thus with a series of 1000 observations, κ will be 6·6 and $\sqrt{\kappa}$ will be about 2·57, and an amplitude in excess of $2 \cdot 57 \times \dfrac{2\sigma}{\sqrt{n-1}}$ will be regarded as possibly significant. Whether the corresponding period is of physical significance cannot be decided from purely statistical argument, and consideration must be given to the possible existence of the same period in other records, and to its persistence in different parts of the series of observations under examination.

The results derived in § 84 may be applied to estimate the extent to which the variability of the observations can be accounted for by the computed amplitudes. The fraction $\dfrac{\Sigma R^2}{2\sigma^2}$ is a measure of the part of the variability of the observations which is so accounted for.

100. Schuster's Investigations.

The periodogram in its original form was developed by Schuster in a paper entitled* "The Investigation of Hidden Periodicities." In this paper the ordinate of the periodogram corresponding to

* *Terrestrial Magnetism*, Vol. III, p. 22.

a period T was defined as the amplitude derived for that period. With our previous notation, $S = R$.

In a later paper*, Schuster gave a further analytical development of the periodogram, taking the ordinate to be the square of the amplitude, or $S = R^2$.

In a third paper†, Schuster gave a detailed development of the analogy between the periodogram and the distribution of energy in a bright lined spectrum produced by a simple grating. The ordinate S of the periodogram was there defined as $\frac{1}{4}R^2 pn\alpha$, using the notation of the present chapter.

Schuster applied the method of the third paper to the discussion of the periodicities of sunspots‡. In the course of this work the material accumulated during 150 years was used, and the periodogram was drawn for values of T varying from 55 days to 24 years. The whole range of 150 years was then divided into two parts, of 75 years each, and the periodogram for each of these parts drawn separately. A very striking result was obtained. The two curves bore not the slightest resemblance to each other. During the first interval of 75 years the principal periods were $13\frac{3}{4}$ years and $9\frac{1}{4}$ years, while during the second interval, up to 1900, the $11\frac{1}{8}$ year period predominated. A number of other fairly clearly-defined periods were also discussed.

The whole discussion points to the conclusion that in such observational data as are generally treated by the periodogram method, we meet with a new type of periodicity which is ill-defined, and not always persistent in amplitude or phase. Outside the domain of gravitational astronomy one seldom finds isolated and clearly-defined periodicities. In the periodogram the problem is often complicated by the presence of not merely one or two well-marked periods, but of a large number of almost coincident periods, so that the periodogram may then be compared to a spectrum full of closely-crowded lines. In such a case, a high resolving power is necessary in order to separate out the slightly differing periods; i.e. the observations used in the discussion must extend over as long a period as possible.

* *Camb. Phil. Soc. Trans.*, Vol. xviii, p. 107.
† *Proc. Roy. Soc.*, Vol. lxxvii, p. 136.
‡ *Phil. Trans. Roy. Soc.*, 206 A, pp. 69—100.

101. Whittaker's variant of Schuster's method.

In a discussion of the variations of SS Cygni*, Gibb applies a method suggested by Whittaker, the essential point of which is to defer the harmonic analysis until a late stage of the work. To examine the observed quantities u_0, u_1, u_2, etc. for a period of p intervals we arrange them as before in n rows of p each, and add the columns, forming the sums V_0, V_1, V_2, etc. Then, instead of applying Fourier's analysis to the V's, we find the difference between the greatest and least of these quantities. This is a rough measure of the amplitude of the period of p intervals.

Let y be the difference between the greatest and the least of the V's for a trial period x (where $x = p\alpha$). Draw a graph with x as abscissa and y as ordinate. The resulting curve may be called the "curve of periods." It shows high peaks at the points at which real periodicities exist, and lower peaks at points corresponding to doubtful periodicities. Such a curve of periods indicates the range of possible periods which can be usefully explored by a detailed periodogram analysis, and forms a useful preliminary to the application of Schuster's method.

102. Use of the Complete Fourier Sequence.

Turner† has suggested that the work of the periodogram might be more economically performed by the evaluation of the complete Fourier sequence of exact sub-multiples of the whole range of time covered by the observations. When the figures analysed form a regular series, with no sudden jumps between two consecutive values, it may be expected that the later coefficients in the Fourier sequence will be small. The observations can then be completely represented (with considerable accuracy) by a limited number of terms, and, theoretically at any rate, the Fourier sequence should contain within itself all the properties of the original observations, including the periodicities of those observations.

In the first place we must consider the effect of a period of the original observations whose length falls between two consecutive sub-multiples of the total range of time covered by the observations,

---

* *Monthly Notices, R.A.S.*, Vol. LXXIV, p. 678.
† *Monthly Notices, R.A.S.*, Vol. LXXIII, pp. 549, 714.

and so falls between the periods represented by two consecutive terms of the Fourier sequence. It might be anticipated *a priori* that such a true period should affect most those terms in the Fourier sequence whose periods most nearly coincide with its own. This can easily be proved by simple calculation. Let the true period be represented by a single term, with unit amplitude

$$\sin{(pt + \delta)}.$$

Let the resulting Fourier sequence be

$$\sum_{q=0} A_q \cos + \sum_{q=1} B_q \sin qt.$$

It will be assumed that the number of observations is sufficiently large to permit of our replacing summations by integrations. This simplifies the work considerably, and does not in any way affect the conclusions arrived at. We then have

$$\pi B_q = \int_0^{2\pi} \sin{(pt + \delta)} \sin qt \, dt$$

$$= \frac{2q}{p^2 - q^2} \sin \pi p \cos{(\pi p + \delta)},$$

$$\pi A_q = \int_0^{2\pi} \sin{(pt + \delta)} \cos qt \, dt$$

$$= \frac{2p}{p^2 - q^2} \sin \pi p \sin{(\pi p + \delta)}.$$

If p lies between q and $q + 1$, it is clear that $p^2 - q^2$, and consequently A and B, change sign as we go from q to $q + 1$. This change of sign affords a new criterion for evaluating periodicities, but is strictly valid only when the periodicity considered is isolated. Again the values of A and B will be greatest for those values of q which yield the least values of $p^2 - q^2$. Thus if there be an isolated periodicity between q and $q + 1$, the coefficients $A_q, A_{q+1}, B_q, B_{q+1}$ will be greater than any of the other coefficients in their neighbourhood.

Let
$$p = q + x.$$

Then
$$\pi A_q = \frac{2p}{p+q}\frac{\sin \pi x}{x}\sin(\pi x + \delta),$$

$$\pi A_{q+1} = \frac{-2p}{p+q+1}\frac{\sin \pi x}{1-x}\sin(\pi x + \delta),$$

and
$$\frac{A_q}{A_{q+1}} = -\frac{1-x}{x} = \frac{B_q}{B_{q+1}},$$

when q is large. Either of the two expressions will yield the value of x. This value is derived on the assumption that q is large, but this condition will generally hold for any period worthy of serious consideration. If R_q be the amplitude of the qth period in the Fourier sequence,

$$R_q^2 = A_q^2 + B_q^2 \quad \text{and} \quad R_q = \frac{\sin \pi x}{\pi x}, \text{ approximately.}$$

This result may be compared with the equation $\dfrac{R}{R'} = \dfrac{\sin \gamma}{\gamma}$ of § 96.

The following table, extracted from Turner's *Tables for Harmonic Analysis*, shows the effect of a periodic term of unit amplitude, for which $p = q + x$, on the coefficients of the neighbouring terms in the series.

Divisor	$q-2$	$q-1$	q	$q+1$	$q+2$	$q+3$
Factor	$+\dfrac{\sin \pi x}{\pi(x+2)}$	$+\dfrac{\sin \pi x}{\pi(x+1)}$	$+\dfrac{\sin \pi x}{\pi x}$	$-\dfrac{\sin \pi x}{\pi(1-x)}$	$-\dfrac{\sin \pi x}{\pi(2-x)}$	$-\dfrac{\sin \pi x}{\pi(3-x)}$
$x=0$	·00	·00	1·00	·00	·00	·00
$x=\frac{1}{6}$	·07	·14	·95	−·19	−·09	−·06
$x=\frac{1}{3}$	·12	·21	·82	−·41	−·16	−·10
$x=\frac{1}{2}$	·13	·21	·63	−·63	−·21	−·13
$x=\frac{2}{3}$	·10	·16	·41	−·82	−·21	−·12
$x=\frac{5}{6}$	·06	·09	·19	−·95	−·14	−·07
$x=1$	·00	·00	·00	−1·00	·00	·00

This table gives the value of the factor $\dfrac{\sin \pi p}{\pi (p - q)}$ in the values

for A_q and B_q given above, the factor $\dfrac{2p}{p + q}$ or $\dfrac{2q}{p + q}$ being treated as unity.

If we consider the amplitudes

$$\cdots \; \frac{\sin \pi x}{\pi (x + 2)}, \; \frac{\sin \pi x}{\pi (x + 1)}, \; \frac{\sin \pi x}{\pi x}, \; -\frac{\sin \pi x}{\pi (1 - x)}, \; -\frac{\sin \pi x}{\pi (2 - x)}, \cdots$$

it can be readily shown that the sum of the squares of all the terms in the series is unity, by using the well-known algebraic series

$$\frac{\pi^2}{\sin^2 \pi x} = \sum_{n = -\infty}^{n = +\infty} \frac{1}{(n + x)^2}.$$

Thus if any one periodic term of unit amplitude be represented by a complete Fourier sequence, the sum of the squares of all the computed amplitudes will be unity. The only Fourier coefficients which have appreciable magnitude are those whose periods are near to that of the original variation. It is clear that if any true period exists with amplitude R, at least one amplitude in the Fourier sequence will have a magnitude as great as $\cdot 63 R$. In using the Fourier sequence it is necessary to bear this in mind and, in the first examination of the terms, to take into consideration amplitudes which are as low as $\cdot 63$ times the amplitude which is regarded as significant.

It is clear from the table that a single periodicity can only yield large values to a few coefficients in the Fourier series. This agrees with the hump in the periodogram which occurs in the neighbourhood of a true period. Thus the Fourier series ought in theory to yield the values of the true periodicities with less work than Schuster's method demands. For the occurrence of two or three large coefficients in the Fourier sequence indicates the presence of a true period between the two terms which have the largest coefficients, and the equation

$$\frac{1 - x}{x} = -\frac{A_q}{A_{q+1}} = -\frac{B_q}{B_{q+1}}.$$

should give the accurate value of x, replacing the secondary analysis of Schuster's method.

In practice, however, the matter is by no means so simple, as may be seen by considering the following extract from Turner's table of the Fourier sequence for Wolf's sunspot numbers (156 years).

Divisor	Period	A	B	R
12	13·0 years	4·8	5·9	7·6
13	12·0	3·1	14·7	15·0
14	11·14	15·8	−19·6	25·8
15	10·40	−4·8	−13·2	14·0
16	9·75	13·6	0·4	13·6
17	9·20	3·7	6·5	7·5

The largest value of R is at $q = 14$, showing the existence of a true period near 11·14 years, near the supposedly well-known 11 year period. According to the theory given above there should be a change of sign of both A and B on one side of $q = 14$, but not on the other. But the table shows a change of sign of A from $q = 14$ to $q = 15$, and of B from $q = 13$ to $q = 14$. It is thus probable that more than one real periodicity falls within the range of the table given above.

When the changes of sign of A and B agree, it will generally be found that the values of x deduced from the A's and the B's differ. In such a case it is perhaps better to use the equation

$$- \frac{1 - x}{x} = \frac{A_q \pm B_q}{A_{q+1} \pm B_{q+1}}.$$

Take, for example, the following extract from the Fourier sequence for the Greenwich temperatures:

q	A	B	R
22	·4	1·0	1·0
23	−3·0	4·3	5·2
24	·7	−2·8	2·9

Putting
$$-\frac{1-x}{x} = \frac{7\cdot3}{3\cdot5},$$

we find
$$x = \frac{3\cdot5}{10\cdot8} = \cdot324;$$

therefore　　　　　$p = 23\cdot324.$

When a large number of consecutive terms of the Fourier sequence are large, or the law of signs of A and B is not followed, it is generally safe to conclude that a number of periodicities are involved. In such a case it is best to examine the observations for possible discontinuities.

103. The Investigation of Discontinuities *.

In order to investigate the possible discontinuities of the 11 year sunspot period, Turner arranged the sunspot numbers in periods of 12 years. The coefficients A, B were evaluated for years 1 to 12. Then the year 13 was substituted for the year 1, and the coefficients A, B were re-calculated. Next the year 14 was substituted for the year 2; and the process was continued until the whole series of sunspot numbers had been used up. The work can be carried out in the following way. In Table I are given the Wolf sunspot numbers arranged in periods of 12 years. In Table II the first row is identical with the first row in Table I, the second row is the difference of row 2 and row 1 of Table I, etc.

TABLE I.

Year												
1749	81	83	48	48	31	12	10	10	32	48	54	63
1761	86	61	45	36	21	11	38	70	106	101	82	66
1773	35	31	7	20	92	154	126	85	68	38	23	10
etc.					etc.							

TABLE II.

1749	81	83	48	48	31	12	10	10	32	48	54	63
1761	5	−22	− 3	−12	−10	−1	28	60	74	53	28	3
1773	−51	−30	−38	−16	71	143	88	15	−38	−63	−59	−56
etc.					etc.							

TABLE III.
Products for cosines.

1749	81	72	24	0	−16	−10	−10	− 9	−16	0	27	55
1761	5	−19	− 2	0	5	1	−28	−52	−37	0	14	3
1773	−51	−26	−19	0	−36	−124	−88	−13	19	0	−30	−48

* Turner, *Monthly Notices, R.A.S.*, Vol. LXXIV, p. 82.

TABLE IV.

Products for sines.

Year											
1749	0	42	41	48	27	6	0	-5	-28	-48	-47	-32
1761	0	-11	-3	-12	-9	0	0	-30	-65	-53	-24	-2
1773	0	-15	-33	-16	62	72	0	-8	33	63	51	28

Table III is derived by multiplying the columns of Table II by the appropriate factors cos 0°, cos 30°, cos 60°, etc. The factors are 1, ·866, ·5, 0, $-$·5, $-$·866, -1, $-$·866, $-$·5, 0, ·5, ·866. Similarly Table IV is derived by multiplying the columns of Table II by the appropriate factors sin 0°, sin 30°, sin 60°, etc. These factors are 0, ·5, ·866, 1, ·866, ·5, 0, $-$·5, $-$·866, -1, $-$·866, $-$·5.

The difference between A, B for the 12 year interval starting at 1749, and the interval starting at 1750, is due to the substitution of 86 for 81 in the first term. The difference, 5, is given in Table II, and is multiplied by the appropriate factors, 1 and 0, in Tables III and IV. To find A, B, for the 12 year interval starting at 1751, we must substitute 61 for 83 in the second term. This is the difference of -22 given in Table II, and multiplied by the appropriate factors in Tables III and IV. The process of evaluation of successive A's and B's is thus reduced to the addition of successive terms of Tables III and IV.

Thus we obtain a series of values of A and B. The phase ϕ is deduced from the formula

$$\tan \phi = \frac{B}{A}.$$

From the progression of the values of ϕ the accurate period can be evaluated as from fig. 13. In the work on the sunspot numbers, Turner found that the progression of phase underwent sudden changes near the years 1766, 1796, 1838, 1868, and 1895, while in the intervals between these breaks the progression of phase was regular. Thus from 1749 to 1766 the phase could be represented by a formula

$$\phi = \text{const.} - 3° . t,$$

where t is measured in years. This corresponds to a period of

$$\frac{12}{1 + \frac{36}{360}} \text{ years} = 10\cdot9 \text{ years}.$$

Again from about 1760 to 1796 the phase could be represented with fair accuracy by a formula

$$\phi = \text{const.} - 8\cdot3^\circ \cdot t,$$

corresponding to a period of

$$\frac{12}{1 + \frac{12 \times 8\cdot3}{360}} \text{ years, or } 9\cdot37 \text{ years.}$$

Turner's results show clearly that the predominating periodicity changes at intervals, bearing out the results previously derived by Schuster.

Turner's method of investigating discontinuities in the main periodicities is to be recommended in all periodogram work. It affords the surest method of detecting changes in the relative importance of different periodicities from time to time. The labour involved is not prohibitive, particularly when it is considered in conjunction with the importance of its results.

104. An Application of the Fourier Sequence in Periodogram Work.

The following brief details of an analysis of the Greenwich Temperature records will illustrate the method of using the complete Fourier sequence. The records used in the first analysis extended from 1841 to 1890, and the monthly mean temperatures were used. The effect of the annual period and all its submultiples was removed from the records by subtracting from each monthly mean the mean temperature of the corresponding month during the whole interval of 50 years. These quantities subtracted were analysed separately (see Ex. 5, page 196). In order to avoid negative signs as far as possible, 10° was added to the figures obtained after subtracting the general mean for the whole period from the monthly means. The figures so obtained, expressed in units of $\cdot1^\circ$, were taken as the material for the periodogram analysis. There were $12 \times 50 = 600$ figures.

Adding together the figures for 25 consecutive months, and dividing by 25, we obtain 24 terms, each of which is the mean deviation for 25 months. These figures were analysed by the usual method up to the 4th harmonic (see Ex. 6, p. 197). Next, the figures for 10 consecutive months were added, yielding 60 terms, and these were analysed for the 5th, 6th,...up to the 14th harmonics. The first 14 periods of the Fourier sequence were thus obtained, except that they needed a correction of phase, the two sets of terms being determined with regard to different origins of time.

Certain sets of coefficients were capable of fairly easy evaluation. Thus the period of 40 months and its submultiples were obtained by writing down in a table the 15 periods of 40 months, and taking the means of the columns. The analysis of these means yielded the terms $q = 15$, 30, 45, 60, etc. in the Fourier sequence.

The terms $q = 20$, 40, 60, etc., and the terms $q = 25$, 75, etc., were also evaluated in a similar manner. The remainder of the terms up to $q = 49$ were calculated by a straightforward application of the formulae of § 85, though this involved considerable labour.

The resulting Fourier sequence shows a number of fairly clearly marked periodicities, but a closer examination reveals evidences of discontinuities *.

The figures obtained, when the general mean for 50 years for any one month is subtracted from the corresponding monthly means, do not form a regular series as do the sunspot numbers. It cannot be expected that the coefficients of the later harmonics in the series will become vanishingly small. The apparent chance distribution of the figures analysed will tend to give fairly large values of the later coefficients even when there are no real periodicities in that region. In other words, the mean value of R^2 will be large in regions where no real periodicity exists, and so it will be necessary to consider very carefully whether the apparent periods obtained may not be due to a chance distribution of the figures analysed.

105. Brooks' Difference Periodogram.

A method which defers the harmonic analysis until a late stage of the work is that described by C. E. P. Brooks as the Difference Periodogram†. The method is as follows:

(1) Divide the series into a number of equal sections each of length U. Let the mean value for successive sections be a, b, c, d, etc.

* Vide *Q. J. Roy. Met. Soc.*, XLV, p. 323, 1919.
† *Proc. Roy. Soc.*, A, Vol. 105, pp. 346—359, 1924.

(2) Form the differences $(a-b)$, $-(b-c)$, $(c-d)$, $-(d-e)$, etc.

(3) Form the successive sums of the quantities in (2), i.e. $\{(a-b)-(b-c)\}$, $\{-(b-c)+(c-d)\}$, etc.

(4) Plot the figures obtained in (3) with a horizontal time scale.

It will then be found that if there is a true period near $2U$ in the original observations, a period of length C will appear in the graph (4), and the length of the period T in the original observations is given by

$$\frac{1}{T} - \frac{1}{2U} = \pm \frac{1}{C} \quad \text{or} \quad T = 2U\,\frac{C}{C \pm 2U}.$$

For a given C, there are thus two possible values of T. To determine which is the appropriate value, the amplitudes of the two alternatives may be evaluated by harmonic analysis, or the difference periodogram may be worked out for a slightly different value of U.

If the existence of a period between certain limits x and y is suspected, the difference periodogram should be worked out for a convenient value of U either less than $\frac{1}{2}x$ or greater than $\frac{1}{2}y$. Then it will follow that one of the values of T will fall within the region between x and y, while the other falls outside that range.

If the curve (4) is very irregular, we may modify stage (3), and, instead of taking the sums of successive quantities formed in (2), take the means or sums of three or four of these quantities.

The work involved in this method is simple and straightforward up to the stage of interpretation of the curves (4). If there is a well-defined period in the region investigated, the curve will be easy to interpret, but occasionally it is found that these curves are almost as complicated as the graphs of the original data.

106. Other Methods of investigating Periodicities.

A method depending on the formation of successive differences was given by Lagrange (*Œuvres*, VI, p. 505, VII, p. 535). This method was improved by Dale*, but was not applied to the detailed investigation of a series of observations.

* *Monthly Notices, R.A.S.*, Vol. 74, p. 628, 1914.

Whitaker and Robinson* suggested a new method of attack by the evaluation of the standard deviation of the V's of §87, and comparing its square with the square of the standard deviation of the original observations. The method has the defect that it does not distinguish between a trial period and its sub-harmonics.

It has been frequently suggested that if the correlation coefficient between $l_0, l_1, l_2, \ldots$ etc. and $l_t, l_{t+1}, l_{t+2}, \ldots$ etc. be evaluated for different values of t, the correlation coefficients will show maxima at the values of t corresponding to true periods in the observations. The method would be somewhat laborious if the correlation coefficients were all evaluated by straightforward arithmetic.

Craig† proposed to calculate the correlation coefficients between the original observations and the series

$$1, \quad \cos \theta, \quad \cos 2\theta, \ldots$$

$$\sin \theta, \quad \sin 2\theta, \ldots$$

where θ takes the value $\dfrac{2\pi}{p}$ in the investigation of a trial period of p intervals.

Bush, Gauge and Stuart‡ have described a "continuous integraph," an electrical device for evaluating the integral of the product of any two continuous functions. It will plot $F(x)$ against x, where

$$F(x) = \int_a^x f(x)\, g(x)\, dx.$$

By making $g(x) = f(x+t)$, such an instrument could be made to yield the correlation coefficients between the series l_0, l_1, l_2, etc. and the series l_t, l_{t+1}, l_{t+2}, etc. very readily. Further, by making $g(x)$ equal to $\cos x$ or $\sin x$, we could evaluate the amplitudes and phases of the trial periods which had been shown to be of importance, or carry out a complete investigation along the lines suggested by Craig.

* *Calculus of Observations*, p. 346.

† London, *Rep. Brit. Ass.*, 1912, p. 416.

‡ *Journal Franklin Institute*, January, 1927.

Yule* has given a method of investigating periodicities in disturbed series, which, while not giving the most rapid method of evaluating a single period, should be of great value in discussing the physical nature of variations in the observations. Foster† has given an optical method of determining periodicities from the graph of observations.

* *Phil. Trans. Roy. Soc.*, 226 A, p. 267, 1927.
† *Journal Textile Institute*, January, 1930.

APPENDIX I

LIST OF REFERENCES

The following short list of references, supplementing those given in the text, may prove of use to the student who desires to read original memoirs bearing upon different parts of the subject.

CHAPTER II

BERNSTEIN. "On Gauss's Error Law." *Math. Annalen*, Vol. LXIV, p. 417.

GLAISHER. "On the Law of Facility of Errors of Observations, and on the Method of Least Squares." *Mem. R.A.S.*, Vol. XXXIX. (A general discussion of a number of different proofs of Gauss's Law of Errors.)

SEARLE. "Geometrical Methods of Combining Observations." *Lick Observatory Bulletin*, Vol. LX.

STORY. "The Law of Error developed from a new Standpoint." *Proc. Amer. Acad. Arts and Sci.*, Vol. LX, p. 167.

Full Tables of $\dfrac{2}{\sqrt{\pi}} \displaystyle\int_0^t e^{-t^2} dt$. *Trans. Royal Soc. Edin.*, Vol. XXXIX, p. 257.

CHAPTER VI

GLAISHER. "Determinant Method of Solution of the Normal Equations." *Monthly Notices, R.A.S.*, Vol. XL, p. 600; Vol. XLI, p. 18.

—— "Solution of the Normal Equations by Gauss's Method." *Monthly Notices, R.A.S.*, Vol. XXXIV, p. 311.

CHAPTER IX

PEARSON, K. "Skew Variation in Homogeneous Material." *Phil. Trans. R.S.*, Vol. 186 A, p. 343; Vol. 197 A, p. 443.

—— "On the Systematic Fitting of Curves to Observations and Measurements." *Biometrika*, Vol. I, p. 265; Vol. II, p. 1.

—— "On the Curves which are most suitable for describing the Frequency of Random Samples of a Population." *Biometrika*, Vol. V, p. 172.

CHAPTER X

BROWN, W. and THOMSON, G. H. *The Essentials of Mental Measurement.* 3rd edn. Cambridge, 1925. (Applications of Correlation in Psychology)

BURT, C. "Experimental Tests of General Intelligence." *Brit. Jour. Psych.,* Vol. III, p. 94.

ELDERTON, W. P. *Frequency Curves and Correlation.* (Deals largely with actuarial applications.)

FISHER, R. A. *Statistical Methods for Research Workers.* 2nd edn., 1928.

PEARSON, K. "Regression, Heredity, and Panmixia." *Phil. Trans. R.S.,* Vol. 187 A, p. 253.

—— "On the Theory of Contingency and its Relation to Association and Normal Correlation." *Drapers' Co. Research Memoirs,* Biometric Series I, 1904.

—— "On the Theory of Skew-Correlation and Non-Linear Regression." *Drapers' Co. Research Memoirs,* Biometric Series II, 1905.

—— "On Further Methods of Determining Correlation." *Drapers' Co. Research Memoirs,* Biometric Series IV, 1907.

YULE, G. U. *An Introduction to the Theory of Statistics.* 1912.

APPENDIX II

TABLE OF Erf $(t) = \Theta(t) = \dfrac{2}{\sqrt{\pi}} \displaystyle\int_0^t e^{-t^2}\, dt.$

t	$\Theta(t)$	Diff.	t	$\Theta(t)$	Diff.
0·00	0·00000		0·30	0·32863	
·01	·01128	1128	·31	·33891	1028
·02	·02256	1128	·32	·34913	1022
·03	·03384	1128	·33	·35928	1015
·04	·04511	1127	·34	·36936	1008
		1126			1002
0·05	0·05637		0·35	0·37938	
·06	·06762	1125	·36	·38933	995
·07	·07886	1124	·37	·39921	988
·08	·09008	1122	·38	·40901	980
·09	·10128	1120	·39	·41874	973
		1118			965
0·10	0·11246		0·40	0·42839	
·11	·12362	1116	·41	·43797	958
·12	·13476	1114	·42	·44747	950
·13	·14587	1111	·43	·45689	942
·14	·15695	1108	·44	·46623	934
		1105			925
0·15	0·16800		0·45	0·47548	
·16	·17901	1101	·46	·48466	918
·17	·18999	1098	·47	·49375	909
·18	·20094	1095	·48	·50275	900
·19	·21184	1090	·49	·51167	892
		1086			883
0·20	0·22270		0·50	0·52050	
·21	·23352	1082	·51	·52924	874
·22	·24430	1078	·52	·53790	866
·23	·25502	1072	·53	·54646	856
·24	·26570	1068	·54	·55494	848
		1063			838
0·25	0·27633		0·55	0·56332	
·26	·28690	1057	·56	·57162	830
·27	·29742	1052	·57	·57982	820
·28	·30788	1046	·58	·58792	810
·29	·31828	1040	·59	·59594	802
		1035			792
0·30	0·32863		0·60	0·60386	

t	$\Theta(t)$	Diff.	t	$\Theta(t)$	Diff.
0·60	0·60386	782	1·00	0·84270	411
·61	·61168	773	·01	·84681	403
·62	·61941	764	·02	·85084	394
·63	·62705	754	·03	·85478	387
·64	·63459	744	·04	·85865	379
0·65	0·64203	735	1·05	0·86244	370
·66	·64938	725	·06	·86614	363
·67	·65663	715	·07	·86977	356
·68	·66378	706	·08	·87333	347
·69	·67084	696	·09	·87680	341
0·70	0·67780	687	1·10	0·88021	332
·71	·68467	676	·11	·88353	326
·72	·69143	667	·12	·88679	318
·73	·69810	658	·13	·88997	311
·74	·70468	648	·14	·89308	304
0·75	0·71116	638	1·15	0·89612	298
·76	·71754	628	·16	·89910	290
·77	·72382	619	·17	·90200	284
·78	·73001	609	·18	·90484	277
·79	·73610	600	·19	·90761	270
0·80	0·74210	590	1·20	0·91031	265
·81	·74800	581	·21	·91296	257
·82	·75381	571	·22	·91553	252
·83	·75952	562	·23	·91805	246
·84	·76514	553	·24	·92051	239
0·85	0·77067	543	1·25	0·92290	234
·86	·77610	534	·26	·92524	227
·87	·78144	525	·27	·92751	222
·88	·78669	515	·28	·92973	217
·89	·79184	507	·29	·93190	211
0·90	0·79691	497	1·30	0·93401	205
·91	·80188	489	·31	·93606	201
·92	·80677	479	·32	·93807	195
·93	·81156	471	·33	·94002	189
·94	·81627	462	·34	·94191	185
0·95	0·82089	453	1·35	0·94376	180
·96	·82542	445	·36	·94556	175
·97	·82987	436	·37	·94731	171
·98	·83423	428	·38	·94902	165
·99	·83851	419	·39	·95067	162
1·00	0·84270		1·40	0·95229	

t	$\Theta(t)$	Diff.	t	$\Theta(t)$	Diff.
1·40	0·95229		1·80	0·98909	
		156			43
·41	·95385		·81	·98952	
		153			42
·42	·95538		·82	·98994	
		148			41
·43	·95686		·83	·99035	
		144			39
·44	·95830		·84	·99074	
		140			37
1·45	0·95970		1·85	0·99111	
		135			36
·46	·96105		·86	·99147	
		132			35
·47	·96237		·87	·99182	
		128			34
·48	·96365		·88	·99216	
		125			32
·49	·96490		·89	·99248	
		121			31
1·50	0·96611		1·90	0·99279	
		117			30
·51	·96728		·91	·99309	
		113			29
·52	·96841		·92	·99338	
		111			28
·53	·96952		·93	·99366	
		107			26
·54	·97059		·94	·99392	
		103			26
1·55	0·97162		1·95	0·99418	
		101			25
·56	·97263		·96	·99443	
		97			23
·57	·97360		·97	·99466	
		95			23
·58	·97455		·98	·99489	
		91			22
·59	·97546		·99	·99511	
		89			21
1·60	0·97635		2·00	0·99532	
		86			20
·61	·97721		·01	·99552	
		83			20
·62	·97804		·02	·99572	
		80			19
·63	·97884		·03	·99591	
		78			18
·64	·97962		·04	·99609	
		76			17
1·65	0·98038		2·05	0·99626	
		72			16
·66	·98110		·06	·99642	
		71			16
·67	·98181		·07	·99658	
		68			15
·68	·98249		·08	·99673	
		66			15
·69	·98315		·09	·99688	
		64			14
1·70	0·98379		2·10	0·99702	
		62			13
·71	·98441		·11	·99715	
		59			13
·72	·98500		·12	·99728	
		58			13
·73	·98558		·13	·99741	
		55			12
·74	·98613		·14	·99753	
		54			11
1·75	0·98667		2·15	0·99764	
		52			11
·76	·98719		·16	·99775	
		50			10
·77	·98769		·17	·99785	
		48			10
·78	·98817		·18	·99795	
		47			10
·79	·98864		·19	·99805	
		45			9
1·80	0·98909		2·20	0·99814	

t	$\Theta(t)$	t	$\Theta(t)$
2·20	0·99814	2·60	0·99976
·21	·99822	·65	·99982
·22	·99831	·70	·99987
·23	·99839	·75	·99990
·24	·99846	·80	·99993
2·25	0·99854	2·85	0·99994 4
·26	·99861	·90	·99995 9
·27	·99867	·95	·99997 0
·28	·99874	3·00	·99997 8
·29	·99880	·05	·99998 4
2·30	0·99886	3·10	0·99998 84
·31	·99891	·15	·99999 16
·32	·99897	·20	·99999 40
·33	·99902	·25	·99999 57
·34	·99906	·30	·99999 69
2·35	0·99911	3·35	0·99999 78
·36	·99915	·40	·99999 85
·37	·99920	·45	·99999 89
·38	·99924	·50	·99999 93
·39	·99928	·55	·99999 95
2·40	0·99931	3·60	0·99999 96441
·41	·99935	·70	·99999 98329
·42	·99938	·80	·99999 99230
·43	·99941	·90	·99999 99652
·44	·99944	4·00	·99999 99845
2·45	0·99947	4·10	0·99999 99933
·46	·99950	·20	·99999 99971
·47	·99952	·30	·99999 99988
·48	·99955	·40	·99999 99995
·49	·99957	·50	·99999 99998
2·50	0·99959	4·60	0·99999 99999 2
·51	·99961	·70	·99999 99999 7
·52	·99963	·80	·99999 99999 9
·53	·99965	·90	·99999 99999 95
·54	·99967	5·00	$1 - 1 \cdot 5 \times 10^{-12}$
2·55	0·99969	5·50	$1 - 7 \quad \times 10^{-15}$
·56	·99971	6·00	$1 - 2 \cdot 2 \times 10^{-17}$
·57	·99972		
·58	·99974		
·59	·99975		

INDEX

Milton Keynes UK
Ingram Content Group UK Ltd.
UKHW041522181024
449640UK00009B/134

9 781107 697614